Fredholm Operators
and Einstein Metrics on
Conformally Compact Manifolds

Memoirs
of the
American Mathematical Society

Number 864

Fredholm Operators
and Einstein Metrics on
Conformally Compact Manifolds

John M. Lee

September 2006 • Volume 183 • Number 864 (end of volume) • ISSN 0065-9266

American Mathematical Society
Providence, Rhode Island

2000 *Mathematics Subject Classification.* Primary 53C25; Secondary 58J05, 58J60.

Library of Congress Cataloging-in-Publication Data

Lee, John M., 1950–
  Fredholm operators and Einstein metrics on conformally compact manifolds / John M. Lee.
    p. cm. — (Memoirs of the American Mathematical Society, ISSN 0065-9266 ; no. 864)
  "Volume 183, number 864 (end of volume)."
  Includes bibliographical references.
  ISBN-13: 978-0-8218-3915-7 (alk. paper)
  1. Riemannian manifolds. 2. Fredholm operators. 3. Hyperbolic spaces. 4. Differential equations, Elliptic I. Title. II. Series.

QA3.A57 no. 864
[QA671]
510 s—dc22
[516.3'62]   2006045731

Memoirs of the American Mathematical Society

This journal is devoted entirely to research in pure and applied mathematics.

Subscription information. The 2006 subscription begins with volume 179 and consists of six mailings, each containing one or more numbers. Subscription prices for 2006 are US$624 list, US$499 institutional member. A late charge of 10% of the subscription price will be imposed on orders received from nonmembers after January 1 of the subscription year. Subscribers outside the United States and India must pay a postage surcharge of US$31; subscribers in India must pay a postage surcharge of US$43. Expedited delivery to destinations in North America US$35; elsewhere US$130. Each number may be ordered separately; *please specify number* when ordering an individual number. For prices and titles of recently released numbers, see the New Publications sections of the *Notices of the American Mathematical Society*.

Back number information. For back issues see the *AMS Catalog of Publications*.

Subscriptions and orders should be addressed to the American Mathematical Society, P. O. Box 845904, Boston, MA 02284-5904, USA. *All orders must be accompanied by payment.* Other correspondence should be addressed to 201 Charles Street, Providence, RI 02904-2294, USA.

Copying and reprinting. Individual readers of this publication, and nonprofit libraries acting for them, are permitted to make fair use of the material, such as to copy a chapter for use in teaching or research. Permission is granted to quote brief passages from this publication in reviews, provided the customary acknowledgment of the source is given.

Republication, systematic copying, or multiple reproduction of any material in this publication is permitted only under license from the American Mathematical Society. Requests for such permission should be addressed to the Acquisitions Department, American Mathematical Society, 201 Charles Street, Providence, Rhode Island 02904-2294, USA. Requests can also be made by e-mail to reprint-permission@ams.org.

Memoirs of the American Mathematical Society is published bimonthly (each volume consisting usually of more than one number) by the American Mathematical Society at 201 Charles Street, Providence, RI 02904-2294, USA. Periodicals postage paid at Providence, RI. Postmaster: Send address changes to Memoirs, American Mathematical Society, 201 Charles Street, Providence, RI 02904-2294, USA.

© 2006 by the American Mathematical Society. All rights reserved.
Copyright of this publication reverts to the public domain 28 years
after publication. Contact the AMS for copyright status.
This publication is indexed in *Science Citation Index*®, *SciSearch*®, *Research Alert*®,
CompuMath Citation Index®, *Current Contents*®/*Physical, Chemical & Earth Sciences*.
Printed in the United States of America.

∞ The paper used in this book is acid-free and falls within the guidelines
established to ensure permanence and durability.
Visit the AMS home page at http://www.ams.org/

10 9 8 7 6 5 4 3 2 1   11 10 09 08 07 06

Contents

Chapter 1.	Introduction	1
Chapter 2.	Möbius Coordinates	9
Chapter 3.	Function Spaces	13
Chapter 4.	Elliptic Operators	23
Chapter 5.	Analysis on Hyperbolic Space	35
Chapter 6.	Fredholm Theorems	45
Chapter 7.	Laplace Operators	61
Chapter 8.	Einstein Metrics	73
Bibliography		81

Abstract

The main purpose of this monograph is to give an elementary and self-contained account of the existence of asymptotically hyperbolic Einstein metrics with prescribed conformal infinities sufficiently close to that of a given asymptotically hyperbolic Einstein metric with nonpositive curvature. The proof is based on an elementary derivation of sharp Fredholm theorems for self-adjoint geometric linear elliptic operators on asymptotically hyperbolic manifolds.

Received by the editor February 16, 2004.
2000 *Mathematics Subject Classification.* Primary 53C25; Secondary 58J05, 58J60.
Key words and phrases. asymptotically hyperbolic, Einstein metric, conformally compact, Fredholm operator, nonlinear elliptic partial differential equations.
Research supported in part by National Science Foundation grants DMS-8901493 and DMS-9404107.

CHAPTER 1

Introduction

In 1985, Charles Fefferman and Robin Graham [**28**] introduced a new and powerful approach to the study of local invariants of conformal structures, based on the observation that the group of conformal automorphisms of the n-sphere (the Möbius group) is essentially the same as the group of Lorentz transformations of $(n+2)$-dimensional Minkowski space. Concretely, one sees this by noting that the conformal structure of the n-sphere can be obtained by viewing the sphere as a cross-section of the forward light cone in Minkowski space. Fefferman and Graham attempted to embed an arbitrary conformal n-manifold into an $(n+2)$-dimensional Ricci-flat Lorentz manifold. They showed that such a Lorentz metric (which they called the *ambient metric*) can be constructed by formal power series for any conformal Riemannian metric, to infinite order when n is odd and to order $n/2$ when n is even; and that this formal metric is a conformal invariant of the original conformal structure.

Once this Lorentz metric (or at least its formal power series) is constructed, its pseudo-Riemannian invariants then automatically give conformal invariants of the original conformal structure. Combined with later work of Bailey, Eastwood, and Graham [**11**], this construction produces all local scalar conformal invariants of odd-dimensional conformal Riemannian structures.

In [**29**], Graham and I adapted this idea to the global setting. We began with the observation that there is a third natural realization of the Möbius group, as the set of isometries of the hyperbolic metric on the interior of the unit ball $\mathbb{B}^{n+1} \subset \mathbb{R}^{n+1}$. As noted by Fefferman and Graham in [**28**], given a compact Riemannian manifold $(S, \widehat{g})$, the problem of finding a Ricci-flat ambient Lorentz metric for the conformal structure $[\widehat{g}]$ is equivalent to that of finding an asymptotically hyperbolic Einstein metric g on the interior of an $(n+1)$-dimensional manifold-with-boundary $\overline{M}$ that has S as boundary and $[\widehat{g}]$ as *conformal infinity* in the following sense: For any smooth, positive defining function ρ for $S = \partial M$, the metric $\rho^2 g$ extends continuously to $\overline{M}$ and restricts to a metric in the conformal class $[\widehat{g}]$ on S. This suggests the following natural problem: Given a compact manifold-with-boundary $\overline{M}$ and a conformal structure $[\widehat{g}]$ on ∂M, can one find a complete, asymptotically hyperbolic Einstein metric g on the interior manifold M that has $[\widehat{g}]$ as conformal infinity? Such a metric is said to be *conformally compact*. To the extent that the answer is yes and the resulting Einstein metric is unique up to boundary-fixing diffeomorphisms, one would thereby obtain a correspondence between global conformal invariants of $[\widehat{g}]$ and global Riemannian invariants of g. Graham and I showed in [**29**] that every conformal structure on $\mathbb{S}^n$ sufficiently close to that of the round metric is the conformal infinity of an Einstein metric close to the hyperbolic metric.

In recent years, interest in asymptotically hyperbolic Einstein metrics has risen dramatically, in no small part because of the role they play in physics. In fact, the notion of conformal infinity for a (pseudo-) Riemannian metric was originally introduced by Roger Penrose [51] in order to analyze the behavior of gravitational energy in asymptotically flat space-times. More recently, asymptotically hyperbolic Einstein metrics have begun to play a central role in the "AdS/CFT correspondence" of quantum field theory. This is not the place for a complete survey of the relevant literature, but let me just refer the reader to [**4, 2, 6, 14, 16, 15, 18, 29, 30, 31, 32, 37, 38, 46, 50, 52, 57, 59**].

The principal purpose of this monograph is to give an elementary and self-contained account of the following generalization of the perturbation result of [**29**]. First, a couple of definitions: The *Lichnerowicz Laplacian* (cf. [**13**]) is the operator Δ_L defined on symmetric 2-tensors by $\Delta_L = \nabla^*\nabla + 2\mathring{R}c - 2\mathring{R}m$, where $\mathring{R}c$ and $\mathring{R}m$ are the natural actions of the Ricci and Riemann curvature tensors on symmetric 2-tensors given in coordinates by

$$\begin{aligned}\mathring{R}c(u)_{ij} &= \tfrac{1}{2}(R_{ik}u_j{}^k + R_{jk}u_i{}^k), \\ \mathring{R}m(u)_{ij} &= R_{ikjl}u^{kl}.\end{aligned} \quad (1.1)$$

For any conformal class of Riemannian metrics on ∂M, the *Yamabe invariant* is defined as the infimum of the total scalar curvature $\int_{\partial M} S_{\widehat{g}}\, dV_{\widehat{g}}$ over unit-volume metrics $\widehat{g}$ in the conformal class.

THEOREM A. *Let M be the interior of a smooth, compact, $(n+1)$-dimensional manifold-with-boundary $\overline{M}$, $n \geq 3$, and let h be an Einstein metric on M that is conformally compact of class $C^{l,\beta}$ with $2 \leq l \leq n-1$ and $0 < \beta < 1$. Let ρ be a smooth defining function for ∂M, and let $\widehat{h} = \rho^2 h|_{\partial M}$. Suppose the operator $\Delta_L + 2n$ associated with h has trivial L^2 kernel on the space of trace-free symmetric 2-tensors. Then there is a constant $\varepsilon > 0$ such that for any $C^{l,\beta}$ Riemannian metric $\widehat{g}$ on ∂M with $\|\widehat{g} - \widehat{h}\|_{C^{l,\beta}} < \varepsilon$, there exists an Einstein metric g on M that has $[\widehat{g}]$ as conformal infinity and is conformally compact of class $C^{l,\beta}$. In particular, this is the case if either of the following hypotheses is satisfied:*

(a) *h has nonpositive sectional curvature.*
(b) *The Yamabe invariant of $[\widehat{h}]$ is nonnegative and h has sectional curvatures bounded above by $(n^2 - 8n)/(8n - 8)$.*

An important feature of this result is that the conformal compactification of the Einstein metric has optimal Hölder regularity up to the boundary (in terms of standard Hölder spaces on $\overline{M}$), at least when n is even. The work of Fefferman and Graham showed that generically for n even there will be a $\rho^n \log \rho$ term in the asymptotic expansion of $\overline{g}$ if $g = \rho^{-2}\overline{g}$ is Einstein; thus we cannot expect to find Einstein metrics with C^n conformal compactifications in general. If h has a sufficiently smooth conformal compactification and $\widehat{g}$ is close to $\widehat{h}$ with sufficiently many derivatives, then Theorem A gives an Einstein metric that is conformally compact of class $C^{n-1,\beta}$ for any $0 < \beta < 1$, which is the best Hölder regularity that can be expected when n is even. The results of [**21**] show that any Einstein metric with smooth conformal infinity and C^2 conformal compactification actually has an infinite-order asymptotic expansion in powers of ρ and $\log \rho$, and in fact is smooth when n is odd.

1. INTRODUCTION

A version of Theorem A (but with less boundary regularity of the resulting Einstein metrics) was also proved independently by Olivier Biquard [15] around the same time as this monograph was originally completed. In addition, analogous perturbation results starting with Kähler-Einstein metrics on bounded domains have been proved independently by Biquard [15] and John Roth [54], and for Einstein metrics asymptotic to the quaternionic and octonionic hyperbolic metrics by Biquard [15]. See below for more on this. See also [3, 4] for some recent results by Michael Anderson on existence and uniqueness of asymptotically hyperbolic Einstein metrics on 4-manifolds, and [22, 24] for results by Erwann Delay and Marc Herzlich on the related problem of prescribing the Ricci curvatures of metrics close to the Einstein models.

The basic approach in this monograph is similar to that of [29]. Because the Einstein equation is invariant under the full diffeomorphism group of M, it is not elliptic as it stands. We obtain an elliptic equation by adding a gauge-breaking term: Fixing a conformally compact "reference metric" g_0, let $\mathbf{\Delta}_{gg_0}(\mathrm{Id})$ denote the harmonic map Laplacian of the identity map, considered as a map from (M, g) to (M, g_0), and let δ_g be the divergence operator associated with g. Then, as is by now familiar, the nonlinear equation

$$Q(g, g_0) := Rc_g + ng - \delta_g^*(\mathbf{\Delta}_{gg_0}(\mathrm{Id})) = 0 \tag{1.2}$$

is elliptic as a function of g. Under relatively mild assumptions on g (see [29, Lemma 2.2]), the solutions to (1.2) are exactly those Einstein metrics g such that $\mathbf{\Delta}_{gg_0}(\mathrm{Id}) = 0$, i.e., such that $\mathrm{Id}\colon (M, g) \to (M, g_0)$ is harmonic.

The linearization of the left-hand side of (1.2) with respect to g at a conformally compact Einstein metric h is

$$D_1 Q_{(h,h)} = \tfrac{1}{2}(\Delta_\mathrm{L} + 2n), \tag{1.3}$$

where Δ_L is the Lichnerowicz Laplacian defined above. In [29], we proved that when h is the hyperbolic metric on the unit ball $\mathbb{B}^{n+1}$, $\Delta_\mathrm{L} + 2n$ is an isomorphism between certain weighted Hölder spaces. However, our methods in that paper were insufficient to prove a sharp isomorphism theorem, with the consequence that our boundary regularity results were not optimal, so even in the case of hyperbolic space Theorem A is an improvement on the results of [29].

Much of this monograph is devoted to an elementary proof of some general sharp Fredholm and isomorphism theorems for self-adjoint geometric linear elliptic operators on asymptotically hyperbolic manifolds. These operators are, in particular, *uniformly degenerate*, in the terminology introduced in [29] (with perhaps less smoothness of the coefficients than we insisted on there): A partial differential operator P on a manifold with boundary is said to be uniformly degenerate if in local coordinates $(\theta^1, \ldots, \theta^n, \rho)$ such that $\rho = 0$ defines the boundary, P can be written locally as a a system of partial differential operators that are polynomials in the vector fields $(\rho\partial/\partial\theta^1, \ldots, \rho\partial/\partial\theta^n, \rho\partial/\partial\rho)$ with coefficients that are at least continuous up to the boundary.

To get an idea of what can be expected, consider a simple example: the scalar Laplacian $\Delta = d^*d$ on functions. When applied to a function of ρ alone, this becomes an ordinary differential operator with a regular singular point at $\rho = 0$. From classical ODE theory, we know that a second-order ordinary differential operator L with a regular singular point at 0 has two *characteristic exponents* s_1 and s_2, defined by $L(\rho^{s_i}) = O(\rho^{s_i+1})$; if $s_1 \neq s_2$, the homogeneous equation $Lu = 0$

has two independent solutions u_1, u_2 with $u_i = O(\rho^{s_i})$, and the inhomogeneous equation $Lu = f = O(\rho^s)$ can be solved with $u = O(\rho^s)$ whenever s is not a characteristic exponent.

To see how this generalizes to systems of partial differential operators, let $P\colon C^\infty(M;E) \to C^\infty(M;F)$ be a uniformly degenerate operator of order m acting between (real or complex) tensor bundles E and F, and let s be any complex number. Define the *indicial map* $I_s(P)\colon E|_{\partial M} \to F|_{\partial M}$ by setting

$$I_s(P)\overline{u} := \rho^{-s} P(\rho^s \overline{u})|_{\partial M},$$

where $\overline{u}$ is any C^m section of $E|_{\partial M}$, extended arbitrarily to a C^m section of E near ∂M. It follows easily from the definition of uniformly degenerate operators that $I_s(P)$ is a (pointwise) bundle endomorphism whose coefficients in any local coordinates are polynomials in s with continuous coefficients depending on $\theta \in \partial M$. (In the special case of a single ODE with a regular singular point, $I_s(P)$ is just multiplication by a number depending polynomially on s, called the *indicial polynomial* of the ODE.) We say a number s is a *characteristic exponent* for P if $I_s(P)$ is singular somewhere on ∂M. Just as in the ODE case, one can construct formal series solutions to $Pu = f$ in which the characteristic exponents correspond to powers of ρ whose coefficients are arbitrary.

Now consider a formally self-adjoint operator acting on sections of a tensor bundle E of type $\binom{r_1}{r_2}$ (i.e., covariant rank r_1 and contravariant rank r_2). In this case, the difference $r = r_1 - r_2$, which we call the *weight* of E, is of central importance, and the exponent $n/2 - r$ plays a special role. If $\overline{u}$ is a section of E that is continuous up to the boundary, then $\rho^{n/2-r}\overline{u}$ is just on the borderline of being in L^2 (see Lemma 3.2(c)). If P is formally self-adjoint, the set of characteristic exponents turns out to be symmetric about the line $\operatorname{Re} s = n/2 - r$ (Corollary 4.5). Therefore, we define the *indicial radius* of P to be the smallest real number $R \geq 0$ such that P has a characteristic exponent whose real part is equal to $n/2 - r + R$. A little experimentation leads one to expect that $Pu = f$ should be well-posed roughly when u and f behave like $\rho^s \overline{u}$, where $\overline{u}$ is continuous up to the boundary and $n/2 - r - R < s < n/2 - r + R$, because for lower values of s one expects P to have an infinite-dimensional kernel, and for higher values one expects an infinite-dimensional cokernel. (This will be made precise in Chapter 6.) Therefore, a fundamental necessary condition for all of our Fredholm results will be that P has positive indicial radius.

To control the order of vanishing or singularity of u and f at the boundary, we work in weighted Sobolev and Hölder spaces, with weights given by powers of the defining function ρ. Precise definitions are given in Chapter 3, but the spaces we work in can be roughly defined as follows: The weighted Sobolev space $H^{k,p}_\delta$ is just the space of tensor fields of the form $\rho^\delta u$ for u in the usual intrinsic Sobolev space $H^{k,p}$ (tensor fields with k covariant derivatives in L^p with respect to g); and the weighted Hölder space $C^{k,\alpha}_\delta$ consists of tensor fields of the form $\rho^\delta u$ for u in the usual Hölder space $C^{k,\alpha}$. In each case, the Sobolev and Hölder norms are defined with respect to a fixed conformally compact metric on M.

An elementary criterion for operators to be Fredholm on L^2 is given in the following proposition. Terms used in the statements of these results will be defined in Chapters 3 and 4.

PROPOSITION B. *Let (M,g) be a connected asymptotically hyperbolic $(n+1)$-manifold of class $C^{l,\beta}$, with $n \geq 1$, $l \geq 2$, and $0 \leq \beta < 1$, and let $E \to M$ be a geometric tensor bundle over M. Suppose $P\colon C^\infty(M;E) \to C^\infty(M;E)$ is an elliptic, formally self-adjoint, geometric partial differential operator of order m, $0 < m \leq l$. As an unbounded operator, $P\colon L^2(M;E) \to L^2(M;E)$ is Fredholm if and only if there exist a compact set $K \subset M$ and a positive constant C such that*

$$\|u\|_{L^2} \leq C\|Pu\|_{L^2} \text{ for all } u \in C_c^\infty(M \smallsetminus K; E). \qquad (1.4)$$

The main analytic result we need for Theorem A is the following sharp Fredholm theorem in weighted Sobolev and Hölder spaces for geometric elliptic operators. (See below for references to other proofs of these and similar results using different approaches.)

THEOREM C. *Let (M,g) be a connected asymptotically hyperbolic $(n+1)$-manifold of class $C^{l,\beta}$, with $n \geq 1$, $l \geq 2$, and $0 \leq \beta < 1$, and let $E \to M$ be a geometric tensor bundle over M. Suppose $P\colon C^\infty(M;E) \to C^\infty(M;E)$ is an elliptic, formally self-adjoint, geometric partial differential operator of order m, $0 < m \leq l$, and assume P satisfies (1.4).*

(a) *The indicial radius R of P is positive.*
(b) *If $1 < p < \infty$ and $m \leq k \leq l$, the natural extension*

$$P\colon H^{k,p}_\delta(M;E) \to H^{k-m,p}_\delta(M;E)$$

is Fredholm if and only if $|\delta + n/p - n/2| < R$. In that case, its index is zero, and its kernel is equal to the L^2 kernel of P.

(c) *If $0 < \alpha < 1$ and $m < k + \alpha \leq l + \beta$, the natural extension*

$$P\colon C^{k,\alpha}_\delta(M;E) \to C^{k-m,\alpha}_\delta(M;E)$$

is Fredholm if and only if $|\delta - n/2| < R$. In that case, its index is zero, and its kernel is equal to the L^2 kernel of P.

The statement and proof of this theorem extend easily to operators on spinor bundles when M is a spin manifold. We restrict attention here to the case of tensor bundles mainly for simplicity of exposition.

We define a *Laplace operator* to be a formally self-adjoint second-order geometric operator of the form $\nabla^*\nabla + \mathscr{K}$, where $\mathscr{K}$ is a bundle endomorphism. Of course, the ordinary Laplacian on functions and the covariant Laplacian on tensor fields are obvious examples of Laplace operators, as are the Lichnerowicz Laplacian Δ_L defined above, and the Laplace-Beltrami operator on differential forms by virtue of Bochner's formula.

The most important example to which we will apply these results is the Lichnerowicz Laplacian on symmetric 2-tensors.

PROPOSITION D. *If $c \in \mathbb{R}$, the operator $\Delta_\mathrm{L} + c$ acting on symmetric 2-tensors satisfies the hypotheses of Theorem C if and only if $c > 2n - n^2/4$, in which case the indicial radius of $\Delta_\mathrm{L} + c$ is*

$$R = \sqrt{\frac{n^2}{4} - 2n + c}.$$

The essential L^2 spectrum of Δ_L is $[n^2/4 - 2n, \infty)$.

(This characterization of the essential spectrum of Δ_L has also been proved by E. Delay in [**23**], using some of the ideas from an earlier draft of this monograph.)

Another example is the covariant Laplacian on trace-free symmetric tensors of any rank.

PROPOSITION E. *If $c \in \mathbb{R}$, the operator $\nabla^*\nabla + c$ acting on trace-free covariant symmetric r-tensors satisfies the hypotheses of Theorem C if and only if $c > -r - n^2/4$, in which case the indicial radius of $\nabla^*\nabla + c$ is*
$$R = \sqrt{\frac{n^2}{4} + r + c}.$$
The essential L^2 spectrum of $\nabla^*\nabla$ is $[n^2/4 + r, \infty)$.

Also, for completeness, we point out the following result, originally proved in the L^2 case by Rafe Mazzeo (cf. [**39, 40, 7**]).

PROPOSITION F. *The Laplace-Beltrami operator acting on q-forms satisfies the hypotheses of Theorem C in the following cases:*
  (a) *When $0 \le q < n/2$, with $R = n/2 - q$.*
  (b) *When $n/2 + 1 < q \le n + 1$, with $R = q - n/2 - 1$.*
In each case, the essential L^2 spectrum of Δ is $[R^2, \infty)$.

Finally, we describe one significant non-Laplace operator to which Theorem C applies. The *conformal Killing operator* is the operator $L\colon C^\infty(M;TM) \to C^\infty(M; \Sigma_0^2 M)$ (where $\Sigma_0^2 M$ is the bundle of trace-free symmetric 2-tensors) defined by letting LV be the trace-free part of the symmetrized covariant derivative of V (see Chapter 7 for details). A vector field V satisfies $LV = 0$ if and only if the flow of V preserves the conformal class of g. The operator L^*L, sometimes called the *vector Laplacian*, plays an important role in the conformal approach to constructing initial data for the Einstein field equations of general relativity; see [**8, 33, 49**].

PROPOSITION G. *The vector Laplacian L^*L satisfies the hypotheses of Theorem C, with $R = n/2 + 1$.*

Most of the analytic results in Theorem C and Propositions D–G are not really new. The systematic treatment of elliptic uniformly degenerate operators dates back to the work of Mazzeo, building on earlier work of Richard Melrose and others [**44, 45, 42, 39, 40**], and Theorem C can also be derived from Mazzeo's microlocal "edge calculus" (cf. [**41**, Theorem 6.1]). Also, L^2-Fredholm criteria for a general class of operators including the ones considered here have been obtained by Robert Lauter, Bertrand Monthubert, and Victor Nistor [**36**, Thm. 4], using the theory of pseudodifferential operators on groupoids. For many purposes in geometric analysis, however, it is useful to have a more "low-tech" approach that does not use pseudodifferential operators. An elementary approach to uniformly degenerate operators based on a priori estimates has been used by Graham and Lee [**29**], Lars Andersson and Piotr Chrusciel [**7, 8**], Johan Råde [**53**], and Michael Anderson [**4, 3, 5**].

The exposition I present here is based on this low-tech approach, and consists of three main ingredients: sharp a priori L^2 estimates, a decay estimate for the hyperbolic Green kernel using the spherical symmetry of the ball model, and a technique due to Råde [**53**] for piecing together a parametrix out of this model Green kernel. One advantage of this approach is that it deals quite naturally with operators whose coefficients are not smooth up to the boundary, a feature that is crucial for the application to Einstein metrics, because the metrics around which we linearize generally have only finite boundary regularity. It is worth noting that for many Laplace operators, the Green kernel estimates are needed only for extending the sharp Fredholm results to Hölder and L^p spaces; sharp Fredholm results in weighted L^2 spaces can be obtained by a much more elementary proof based solely on a priori estimates. See Chapter 7 for details.

The main new results in this monograph are the Bochner-type formula (7.7) for Laplacians on tensor-valued differential forms, which yields a completely elementary proof of sharp Fredholm theorems in weighted L^2 spaces and the identification of the essential spectrum (Chapter 7); new sufficient curvature conditions for the invertibility of the Lichnerowicz Laplacian (part (b) of Theorem A); and the construction of Einstein metrics in the optimal $C^{n-1,\beta}$ Hölder class up to the boundary when the conformal infinity is sufficiently smooth.

Theorem C can be viewed as complementary to the regularity results of Andersson and Chruściel; in particular, for an operator of the type considered here, Proposition 6.5 below can be used to obtain sharp "regularity intervals" in the sense defined in [8], and then the results of [8] can be used to derive asymptotic expansions for solutions to equations of the form $Pu = f$ when f and the metric are sufficiently smooth. We leave it to the interested reader to work out the details. See [21] for example, where these ideas play a central role in the proof of boundary regularity for asymptotically hyperbolic Einstein metrics. Similar results can also be obtained using the edge calculus of [41].

The assumption of a weak a priori L^2 estimate (1.4) near the boundary is used primarily to rule out L^2 kernel of the model operator on hyperbolic space. For almost all the examples treated here, this hypothesis is equivalent to positive indicial radius (see Propositions D, E, F, and G above). However, it is possible for (1.4) to fail even when the indicial radius is positive—an example is given by the Laplace-Beltrami operator on differential forms of degree k on a manifold of dimension $2k$, which is not Fredholm despite the fact that its indicial radius is $1/2$ (see [39, 40] and Lemma 7.2 below). Thanks are due to Robin Graham for pointing out the significance of this example.

A word about the history of this monograph is in order. The result of Theorem A was originally announced in preliminary form at an AMS meeting in 1991. Shortly after that meeting, I discovered a gap in the proof, and stopped work on the paper until I found a way to bridge the gap, sometime around 1998. By that time, administrative responsibilities kept me away from writing until early 2001. The first complete version of this monograph was posted on www.arxiv.org in May 2001; the present version is a minor modification of that one.

After this monograph was nearly finished, the beautiful monograph [15] by Olivier Biquard came to my attention. There is substantial overlap between the results of [15] and this monograph—in particular, Biquard proves a perturbation result for Einstein metrics similar to Theorem A, as well as many of the results of Theorem C in the special case of Laplace operators, using methods very similar to those used here. (He also proves much more, extending many of the same results to Einstein metrics that are asymptotic to the complex, quaternionic, and octonionic hyperbolic metrics.) On the other hand, many of the results of this monograph are stronger than those of [**15**], notably the curvature assumptions of in Theorem A and the boundary regularity of the resulting Einstein metrics.

In Chapter 2 of this monograph, we give the main definitions, and describe special "Möbius coordinate charts" on an asymptotically hyperbolic manifold that relate the geometry to that of hyperbolic space. In Chapter 3, we introduce our weighted Sobolev and Hölder spaces and prove some of their basic properties, and in Chapter 4 we prove some basic mapping properties of geometric elliptic operators on these spaces. The core of the analysis begins in Chapter 5, where we undertake a detailed study of the Green kernels for elliptic geometric operators on hyperbolic space. This is then applied in Chapter 6 to construct a parametrix for an arbitrary operator satisfying the hypotheses of Theorem C. Using this analysis, we prove Theorem C and Proposition B. In Chapter 7, we explore how these results apply in detail to Laplace operators, and prove Propositions D, E, F, and G. Finally, in Chapter 8 we construct asymptotic solutions to (1.2) using a delicate procedure that does not lose any boundary regularity, and use these together with Theorem C to prove Theorem A.

Among the many people to whom I am indebted for inspiration and good ideas while this work was in progress, I would particularly like to express my thanks to Lars Andersson, Piotr Chruściel, Robin Graham, Jim Isenberg, Rafe Mazzeo, Dan Pollack, and John Roth. I also would like to apologize to them and to all who expressed interest in this work for the long delay between the first announcement of these results and the appearance of this monograph. Finally, I am indebted to the referee for a number of useful suggestions.

CHAPTER 2

Möbius Coordinates

Let $\overline{M}$ be a smooth, compact, $(n+1)$-dimensional manifold-with-boundary, $n \geq 1$, and M its interior. A *defining function* will mean a function $\rho\colon \overline{M} \to \mathbb{R}$ of class at least C^1 that is positive in M, vanishes on ∂M, and has nonvanishing differential everywhere on ∂M. We choose a fixed smooth defining function ρ once and for all. For any $\varepsilon > 0$, let $A_\varepsilon \subset M$ denote the open subset where $0 < \rho < \varepsilon$.

A Riemannian metric g on M is said to be *conformally compact of class $C^{l,\beta}$* for a nonnegative integer l and $0 \leq \beta < 1$ if for any smooth defining function ρ, the conformally rescaled metric $\rho^2 g$ has a $C^{l,\beta}$ extension, denoted by $\overline{g}$, to a positive definite tensor field on $\overline{M}$. For such a metric g, the induced boundary metric $\widehat{g} := \overline{g}|_{T\partial M}$ is a $C^{l,\beta}$ Riemannian metric on ∂M whose conformal class $[\widehat{g}]$ is independent of the choice of smooth defining function ρ; this conformal class is called the *conformal infinity* of g.

Throughout this monograph, we will use the Einstein summation convention, with Roman indices $i, j, k, \ldots$ running from 1 to $n+1$ and Greek indices $\alpha, \beta, \gamma, \ldots$ running from 1 to n. We indicate components of covariant derivatives of a tensor field by indices preceded by a semicolon, as in $u_{ij;kl}$. In component calculations, we will always assume that a fixed conformally compact metric g has been chosen, and all covariant derivatives and index raising and lowering operations will be understood to be with respect to g unless otherwise specified, except that $\overline{g}^{ij}$ denotes the *inverse* of the metric $\overline{g} = \rho^2 g$, not its raised-index version. Our convention for the components of the curvature tensor is chosen so that the Ricci tensor is given by the contraction $R_{ik} = R_{ijk}{}^j$.

An important fact about conformally compact metrics is that their local geometry near the boundary looks asymptotically very much like that of hyperbolic space. Mazzeo [**39, 40**] showed, for example, that if g is conformally compact of class at least $C^{2,0}$, then g is complete and has sectional curvatures uniformly approaching $-|d\rho|^2_{\overline{g}}$ near ∂M. Thus, if g is conformally compact of class $C^{l,\beta}$ with $l \geq 2$, and $|d\rho|^2_{\overline{g}} = 1$ on ∂M, we say g is *asymptotically hyperbolic of class $C^{l,\beta}$*.

In fact, the relationship with hyperbolic space can be made even more explicit by constructing special coordinate charts near the boundary. Throughout the rest of this chapter, we assume given a fixed metric g on M that is asymptotically hyperbolic of class $C^{l,\beta}$, with $l \geq 2$ and $0 \leq \beta < 1$. Let $\overline{g} = \rho^2 g$, a $C^{l,\beta}$ metric on $\overline{M}$, and let $\widehat{g}$ denote the restriction of $\overline{g}$ to $T\partial M$.

We begin by choosing a covering of a neighborhood of ∂M in $\overline{M}$ by finitely many smooth coordinate charts (Ω, Θ), where each coordinate map Θ is of the form $\Theta = (\theta, \rho) = (\theta^1, \ldots, \theta^n, \rho)$ and extends to a neighborhood of $\overline{\Omega}$ in $\overline{M}$. In keeping with our index convention, we will sometimes denote ρ by θ^{n+1}, and a symbol with a Roman index such as θ^i can refer to any of the coordinates $\theta^1, \ldots, \theta^n, \rho$.

We fix once and for all finitely many such charts covering a neighborhood $\mathscr{W}$ of ∂M in $\overline{M}$. We will call any of these charts "background coordinates" for $\overline{M}$. By compactness, there is a positive number c such that $A_c \subset \mathscr{W}$, and such that every point $p \in A_c$ is contained in a background coordinate chart containing a set of the form

$$\{(\theta, \rho) : |\theta - \theta(p)| < c, 0 \leq \rho < c\}. \tag{2.1}$$

We will use two models of hyperbolic space, depending on context. In the upper half-space model, we regard hyperbolic space as the open upper half-space $\mathbb{H} = \mathbb{H}^{n+1} \subset \mathbb{R}^{n+1}$, with coordinates $(x, y) = (x^1, \ldots, x^n, y)$, and with the hyperbolic metric $\breve{g}$ given in coordinates by $\breve{g} = y^{-2} \sum_i (dx^i)^2$. (As above, x^i can denote any of the coordinates $x^1, \ldots, x^n, x^{n+1} = y$.) The other model is the Poincaré ball model, in which we regard hyperbolic space as the open unit ball $\mathbb{B} = \mathbb{B}^{n+1} \subset \mathbb{R}^{n+1}$, with coordinates $(\xi^1, \ldots, \xi^{n+1})$, and with the hyperbolic metric (still denoted by $\breve{g}$) given by $\breve{g} = 4(1 - |\xi|)^{-2} \sum_i (d\xi^i)^2$, where $|\xi|$ denotes the Euclidean norm.

In this chapter, we will work exclusively with the upper half-space model. For any $r > 0$, we let $B_r \subset \mathbb{H}$ denote the hyperbolic geodesic ball of radius r about the point $(x, y) = (0, 1)$:

$$B_r = \{(x, y) \in \mathbb{H} : d_{\breve{g}}((x, y), (0, 1)) < r\}.$$

It is easy to check by direct computation that

$$B_r \subset \{(x, y) : |x| < \sinh r,\ e^{-r} < y < e^r\},$$

where $|x|$ denotes the Euclidean norm of $x \in \mathbb{R}^n$.

If p_0 is any point in $A_{c/8}$, choose such a background chart containing p_0, and define a map $\Phi_{p_0} : B_2 \to M$, called a *Möbius chart* centered at p_0, by

$$(\theta, \rho) = \Phi_{p_0}(x, y) = (\theta_0 + \rho_0 x, \rho_0 y),$$

where (θ_0, ρ_0) are the background coordinates of p_0. (It is more convenient in this context to consider a "chart" to be a mapping from $\mathbb{H} \subset \mathbb{R}^{n+1}$ into M, rather than from M to $\mathbb{R}^{n+1}$ as is more common.) Because $\rho_0 < c/8$ and $e^2 < 8$, Φ_{p_0} maps B_2 diffeomorphically onto a neighborhood of p_0 in A_c. In these coordinates, p_0 corresponds to the point $(x, y) = (0, 1) \in \mathbb{H}$. For each $0 < r \leq 2$, let $V_r(p_0) \subset A_c$ be the neighborhood of p_0 defined by

$$V_r(p_0) = \Phi_{p_0}(B_r).$$

We also choose finitely many smooth coordinate charts $\Phi_i : B_2 \to M$ such that the sets $\{\Phi_i(B_1)\}$ cover a neighborhood of $M \smallsetminus A_{c/8}$, and such that each chart Φ_i extends smoothly to a neighborhood of $\overline{B}_2$. For consistency, we will also call these "Möbius charts."

The following lemma shows that the geometry of (M, g) is uniformly bounded in Möbius charts.

LEMMA 2.1. *There exists a constant $C > 0$ such that if $\Phi_{p_0} : B_2 \to M$ is any Möbius chart,*

$$\|\Phi_{p_0}^* g - \breve{g}\|_{C^{l,\beta}(B_2)} \leq C, \tag{2.2}$$

$$\sup_{B_2} |(\Phi_{p_0}^* g)^{-1} \breve{g}| \leq C. \tag{2.3}$$

(The Hölder and sup norms in this estimate are the usual norms applied to the components of a tensor in coordinates; since $\overline{B}_2$ is compact, these are equivalent to the intrinsic Hölder and sup norms on tensors with respect to the hyperbolic metric.)

PROOF. The estimate is immediate for the finitely many charts covering the interior of M, so we need only consider Möbius charts near the boundary. In background coordinates, g can be written

$$g = \rho^{-2}\overline{g}_{ij}(\theta,\rho)d\theta^i d\theta^j.$$

Pulling back to $\mathbb{H}$, we obtain

$$\Phi_{p_0}^* g - \breve{g} = (\rho_0 y)^{-2}\overline{g}_{ij}(\theta_0+\rho_0 x, \rho_0 y)d(\rho_0 x^i)\, d(\rho_0 x^j) - y^{-2}\delta_{ij}dx^i\, dx^j$$
$$= y^{-2}(\overline{g}_{ij}(\theta_0+\rho_0 x, \rho_0 y) - \delta_{ij})dx^i dx^j.$$

Since y and δ_{ij} are smooth functions bounded above and below together with all derivatives on $\overline{B}_2$, it suffices to estimate $\overline{g}_{ij}(\theta_0+\rho_0 x, \rho_0 y)$. Our choice of background coordinates ensures that the eigenvalues of $\overline{g}_{ij}$ are uniformly bounded above and below by a global constant. Uniform estimates on the derivatives of $\Phi_{p_0}^* g$ follow from the fact that

$$\partial_{x^{i_1}}\cdots\partial_{x^{i_m}}(\Phi_{p_0}^*\overline{g})_{ij} = \rho_0^m \Phi_{p_0}^*(\partial_{\theta^{i_1}}\cdots\partial_{\theta^{i_m}}\overline{g}_{ij}).$$

Finally, an easy computation yields uniform Hölder estimates for the lth derivatives of $\Phi_{p_0}^* g$. (See Lemma 6.1 below for a sharper estimate.) □

The next lemma is a version of the well-known Whitney covering lemma adapted to the present situation.

LEMMA 2.2. *There exists a countable collection of points $\{p_i\} \subset M$ and corresponding Möbius charts $\Phi_i = \Phi_{p_i}\colon B_2 \to V_2(p_i) \subset M$ such that the sets $\{V_1(p_i)\}$ cover M and the sets $\{V_2(p_i)\}$ are uniformly locally finite: There exists an integer N such that for each i, $V_2(p_i)$ has nontrivial intersection with $V_2(p_j)$ for at most N values of j.*

PROOF. We only need to show that there exist points $\{p_i\} \subset A_{c/8}$ such that $\{V_1(p_i)\}$ cover $A_{c/8}$ and $\{V_2(p_i)\}$ are uniformly locally finite, for then we can choose finitely many additional charts for the interior without disturbing the uniform local finiteness.

By the preceding lemma, there are positive numbers $r_0 < r_1$ such that each set $V_1(p)$ contains the g-geodesic ball of radius r_0 about p, and each set $V_2(p)$ is contained in the geodesic ball of radius r_1. Let $\{p_i\}$ be any maximal collection of points in $A_{c/8}$ such that the open geodesic balls $\{B_{r_0/2}(p_i)\}$ are disjoint. (Such a maximal collection exists by an easy application of Zorn's lemma.) If p is any point in $A_{c/8}$, by the maximality of the set $\{p_i\}$, $B_{r_0/2}(p)$ must intersect at least one of the balls $B_{r_0/2}(p_i)$ nontrivially, which implies that $p \in B_{r_0}(p_i) \subset V_1(p_i)$ by the triangle inequality. Therefore the sets $\{V_1(p_i)\}$ cover $A_{c/8}$. To bound the number of sets $\{V_2(p_i)\}$ that can intersect, it suffices to bound the number of geodesic balls of radius r_1 around points p_i that can intersect. Let i be arbitrary and suppose $B_{r_1}(p_j) \cap B_{r_1}(p_i) \neq \emptyset$ for some j. By the triangle inequality again, $B_{r_0/2}(p_j) \subset B_{2r_1+r_0/2}(p_i)$. Since M has bounded sectional curvature, standard volume comparison theorems (see, for example, [**19**, Theorems 3.7 and 3.9]) yield uniform volume estimates

$$\mathrm{Vol}(B_{r_0/2}(p_j)) \geq C_1, \qquad \mathrm{Vol}(B_{2r_1+r_0/2}(p_j)) \leq C_2.$$

Since the sets $B_{r_0/2}(p_j)$ are disjoint for different values of j, there can be at most C_2/C_1 such points p_j. □

CHAPTER 3

Function Spaces

In this chapter, we will define weighted Sobolev and Hölder spaces of tensor fields that are well adapted to the geometry of asymptotically hyperbolic manifolds. Similar spaces have been defined by other authors; for some examples, see [**8, 7, 22, 24, 29**].

Throughout this chapter, we assume $\overline{M}$ is a connected smooth $(n+1)$-manifold, g is a metric on M that is asymptotically hyperbolic of class $C^{l,\beta}$, with $l \geq 2$ and $0 \leq \beta < 1$, and ρ is a fixed smooth defining function for ∂M. (It is easy to verify that choosing another smooth defining function will replace the norms we define below by equivalent ones, and will leave the function spaces unchanged.)

A *geometric tensor bundle* over $\overline{M}$ is a subbundle E of some tensor bundle $T^{r_1}_{r_2}\overline{M}$ (tensors of covariant rank r_1 and contravariant rank r_2) associated to a direct summand (not necessarily irreducible) of the standard representation of $O(n+1)$ (or $SO(n+1)$ if M is oriented) on tensors of type $\binom{r_1}{r_2}$ over $\mathbb{R}^{n+1}$. We will also use the same symbol E to denote the restriction of this bundle to M. We define the *weight* of such a bundle $E \subset T^{r_1}_{r_2}\overline{M}$ to be $r = r_1 - r_2$. The significance of this definition lies in the way tensor norms scale conformally: With $\bar{g} = \rho^2 g$, an easy computation shows that

$$|T|_g = \rho^r |T|_{\bar{g}} \quad \text{for all } T \in T^{r_1}_{r_2}.$$

We begin by defining Hölder spaces of functions and tensor fields that are continuous up to the boundary. For $0 \leq \alpha < 1$ and k a nonnegative integer, we let $C^{k,\alpha}_{(0)}(\overline{M})$ denote the usual Banach space of functions on $\overline{M}$ with k derivatives that are Hölder continuous of degree α up to the boundary in each background coordinate chart, with the obvious norm. (When $\alpha = 0$, this is just the usual space of functions that are k times continuously differentiable on $\overline{M}$.) If s is a real number such that $0 \leq s \leq k + \alpha$, we define a subspace $C^{k,\alpha}_{(s)}(\overline{M}) \subset C^{k,\alpha}_{(0)}(\overline{M})$ by

$$C^{k,\alpha}_{(s)}(\overline{M}) = \{u \in C^{k,\alpha}_{(0)}(\overline{M}) : u = O(\rho^s)\}.$$

The next lemma gives some elementary properties of these spaces.

LEMMA 3.1. *Suppose $0 \leq \alpha < 1$ and $0 \leq s \leq k + \alpha$.*
 (a) $C^{k,\alpha}_{(s)}(\overline{M}) = \{u \in C^{k,\alpha}_{(0)}(\overline{M}) : \partial^i u / \partial \rho^i|_{\partial M} = 0 \text{ for } 0 \leq i < s\}$.
 (b) $C^{k,\alpha}_{(s)}(\overline{M})$ *is a closed subspace of* $C^{k,\alpha}_{(0)}(\overline{M})$.
 (c) *If j is a positive integer and $j-1 < s \leq j \leq k$, then* $C^{k,\alpha}_{(s)}(\overline{M}) = C^{k,\alpha}_{(j)}(\overline{M})$.
 (d) *If $k < s \leq k + \alpha$, then* $C^{k,\alpha}_{(s)}(\overline{M}) = C^{k,\alpha}_{(k+\alpha)}(\overline{M})$.
 (e) *If $u \in C^{k,\alpha}_{(s)}(\overline{M})$ for $s \geq 1$, then any background coordinate derivative $\partial_i u$ is in $C^{k-1,\alpha}_{(s-1)}(\overline{M})$.*

(f) If δ is a positive real number such that $s + \delta \le k + \alpha$, then $\rho^\delta C^{k,\alpha}_{(s)}(\overline{M}) \subset C^{k,\alpha}_{(s+\delta)}(\overline{M})$.

(g) If j is an integer such that $0 < j \le s$, then $\rho^{-j} C^{k,\alpha}_{(s)}(\overline{M}) \subset C^{k-j,\alpha}_{(s-j)}(\overline{M})$.

PROOF. All of these claims are local, so we fix one background coordinate chart (θ, ρ) and do all of our computations there. Let m be any nonnegative integer less than or equal to k. Applying the one-variable version of Taylor's formula to $u \in C^{k,\alpha}_{(0)}(\overline{M})$, we obtain

$$u(\theta, \rho) = \sum_{i=0}^{m-1} \frac{1}{i!} \rho^i \frac{\partial^i u}{\partial \rho^i}(\theta, 0) + \frac{1}{(m-1)!} \rho^m \int_0^1 (1-t)^{m-1} \frac{\partial^m u}{\partial \rho^m}(\theta, t\rho)\, dt. \qquad (3.1)$$

The integral above is easily shown to define a function of (θ, ρ) that is in $C^{k-m,\alpha}_{(0)}$ up to the boundary and agrees with $\partial^m u / \partial \rho^m$ on ∂M; therefore the last term in (3.1) is $O(\rho^m)$ in general, and if $\alpha > 0$, it is $O(\rho^{m+\alpha})$ if and only if $\partial^m u / \partial \rho^m$ vanishes on ∂M. Part (a) follows easily from this, and (b), (c), (d), and (e) follow from (a). Part (f) is an immediate consequence of the definition, and (g) follows by setting $m = j$ in (3.1) and multiplying through by ρ^{-j}. □

Because of part (b) of this lemma, we can consider $C^{k,\alpha}_{(s)}(\overline{M})$ as a Banach space with the norm inherited from $C^{k,\alpha}_{(0)}(\overline{M})$.

If E is a geometric tensor bundle over $\overline{M}$, we extend this definition to spaces of tensor fields by letting $C^{k,\alpha}_{(s)}(\overline{M}; E)$ be the space of tensor fields whose components in each background coordinate chart are in $C^{k,\alpha}_{(s)}(\overline{M})$. All the claims of Lemma 3.1 extend immediately to tensor fields.

Next we define some spaces of tensor fields over the interior manifold M associated with the asymptotically hyperbolic metric g. Let us start with the Sobolev spaces, for which the definitions are a bit simpler. First, for $1 < p < \infty$ and k a nonnegative integer less than or equal to l, we define $H^{k,p}(M; E)$ to be the usual (intrinsic) L^p Sobolev space determined by the metric g: That is, $H^{k,p}(M; E)$ is the Banach space of all locally integrable sections u of E such that $\nabla^j u$ (interpreted in the distribution sense) is in $L^p(M; E \otimes T^j M)$ for $0 \le j \le k$, with the norm

$$\|u\|_{k,p} = \left(\sum_{j=0}^k \int_M |\nabla^j u|^p\, dV_g \right)^{1/p}.$$

(Here dV_g is the Riemannian density.) In the special case of $H^{0,2}(M; E) = L^2(M; E)$, we will denote the norm $\|\cdot\|_{0,2}$ and its associated inner product simply by $\|\cdot\|$ and $(\cdot, \cdot)$, respectively. (We reserve the notations $|\cdot|_g$ and $\langle \cdot, \cdot \rangle_g$ for the pointwise norm and inner product on tensors.)

For each real number δ, we define the *weighted Sobolev space* $H^{k,p}_\delta(M; E)$ by

$$H^{k,p}_\delta(M; E) := \rho^\delta H^{k,p}(M; E) = \{\rho^\delta u : u \in H^{k,p}(M; E)\}$$

with norm

$$\|u\|_{k,p,\delta} := \|\rho^{-\delta} u\|_{k,p}.$$

These are easily seen to be Banach spaces, and $H^{k,2}_\delta(M; E)$ is a Hilbert space. Note that Hölder's inequality implies that when $1/p + 1/p^* = 1$, $H^{0,p^*}_{-\delta}(M; E)$ is naturally isomorphic to the dual space of $H^{0,p}_\delta(M;E)$ under the L^2 pairing $(u,v) = \int_M \langle u, v \rangle_g \, dV_g$.

The following lemma is elementary but often useful.

LEMMA 3.2. *Let (M,g) be a conformally compact $(n+1)$-manifold, let ρ be a defining function for M, and let u be a continuous section of a natural tensor bundle E of weight r on M.*

(a) *If $|u|_g \leq C\rho^s$ with s real and greater than $\delta + n/p$, then $u \in H^{0,p}_\delta(M;E)$.*
(b) *If $|u|_g \geq C\rho^s > 0$ on the complement of a compact set, with $s \in \mathbb{R}$ and $s \leq \delta + n/p$, then $u \notin H^{0,p}_\delta(M;E)$.*
(c) *If $u = \rho^s \bar{u}$, where $s \in \mathbb{C}$ and $\bar{u}$ is continuous on $\overline{M}$ and does not vanish identically on ∂M, then $u \in H^{0,p}_\delta(M;E)$ if and only if $\operatorname{Re} s > \delta + n/p - r$.*

PROOF. Let $\bar{g} = \rho^2 g$, which is a continuous Riemannian metric on $\overline{M}$. The lemma follows from the easily-verified facts that $|\rho^s \bar{u}|_g = \rho^{\operatorname{Re} s + r} |\bar{u}|_{\bar{g}}$, $dV_g = \rho^{-n-1} dV_{\bar{g}}$, and $\int_M \rho^s \, dV_g < \infty$ if and only if $\operatorname{Re} s > n$. □

Next we turn to the weighted Hölder spaces. To define Hölder norms for tensor fields on a manifold, one is always faced with the problem of comparing values of a tensor field at nearby points in order to make sense of quotients of the form $|u(x) - u(y)|/d(x,y)^\alpha$. There are various intrinsic ways to do this, such as parallel translating $u(x)$ along a geodesic from x to y, or comparing the components of $u(x)$ and $u(y)$ in Riemannian normal coordinates centered at x; on a noncompact manifold, these can yield different spaces depending on the behavior of the metric near infinity. We will adopt a definition in terms of the Möbius coordinates constructed above which, though not obviously intrinsic to the geometry of (M,g), has the advantage that estimates for elliptic operators in these norms follow very easily from standard local elliptic estimates in Möbius coordinates.

Let α be a real number such that $0 \leq \alpha < 1$, and let k be a nonnegative integer such that $k + \alpha \leq l + \beta$. For any tensor field u with locally $C^{k,\alpha}$ coefficients, define the norm $\|u\|_{k,\alpha}$ by

$$\|u\|_{k,\alpha} := \sup_\Phi \|\Phi^* u\|_{C^{k,\alpha}(B_2)}, \tag{3.2}$$

where $\|v\|_{C^{k,\alpha}(B_2)}$ is just the usual Euclidean Hölder norm of the components of v on $B_2 \subset \mathbb{H}$, and the supremum is over all Möbius charts defined on B_2. Let $C^{k,\alpha}(M;E)$ be the space of sections of E for which this norm is finite. (We distinguish notationally between the Hölder and Sobolev norms as follows: A Greek subscript in the second position takes values in the interval $[0,1)$, and the notation $\|u\|_{k,\alpha}$ denotes a Hölder norm; a Roman subscript takes values in $(1,\infty)$, and the notation $\|u\|_{k,p}$ denotes a Sobolev norm. We will avoid the ambiguous cases $\alpha = 1$ and $p = 1$.) The corresponding *weighted Hölder spaces* are defined for $\delta \in \mathbb{R}$ by

$$C^{k,\alpha}_\delta(M;E) := \rho^\delta C^{k,\alpha}(M;E) = \{\rho^\delta u : u \in C^{k,\alpha}(M;E)\}$$

with norms

$$\|u\|_{k,\alpha,\delta} := \|\rho^{-\delta} u\|_{k,\alpha}.$$

If $U \subset M$ is a subset, the restricted norms are denoted by $\|\cdot\|_{k,p,\delta;U}$ and $\|\cdot\|_{k,\alpha,\delta;U}$, and the spaces $H^{k,p}_\delta(U;E)$ and $C^{k,\alpha}_\delta(U;E)$ are the spaces of sections over U for which these norms are finite. When E is the trivial line bundle (i.e. when the tensor fields in question are just scalar functions), we omit the bundle from the notation: For example, $H^{k,p}(M)$ is the Sobolev space of scalar functions on M with k covariant derivatives in $L^p(M)$.

It is obvious from the definitions that $\rho^\delta \in C^{k,\alpha}_\delta(M)$ for every k. More importantly, we have the following lemma.

LEMMA 3.3. *If $0 \leq j \leq l$, then $\nabla^j \rho \in C^{l-j,\beta}_1(M; T^j M)$.*

PROOF. From the definition of $C^{k,\beta}_1(M; T^j M)$, we need to show for any Möbius chart Φ_{p_0} that the coefficients of $\Phi^*_{p_0}(\rho^{-1} \nabla^j \rho)$ are in $C^{l-j,\beta}(B_2)$, with norm bounded independently of p_0. Since $\Phi^*_{p_0} \rho = \rho(p_0) y$, this follows immediately from the coordinate expression for $\Phi^*_{p_0}(\rho^{-1}\nabla^j \rho) = y^{-1}(\nabla_{\Phi^*_{p_0} g})^j y$ and the fact that the Christoffel symbols of $\Phi^*_{p_0} g$ in Möbius coordinates are uniformly bounded in $C^{l-1,\beta}(B_2)$. □

We have defined the weighted spaces by multiplying the standard Sobolev and Hölder spaces by powers of ρ. In many circumstances, it is more convenient to use an alternative characterization in terms of weighted norms of covariant derivatives of u, as given in the following lemma.

LEMMA 3.4. *Let u be a locally integrable section of a tensor bundle E over an open subset $U \subset M$.*

(a) *For $1 < p < \infty$ and $0 \leq k \leq l$, $u \in H^{k,p}_\delta(U; E)$ if and only if $\rho^{-\delta} \nabla^j u \in L^p(U; E \otimes T^j M)$ for $0 \leq j \leq k$, and the $H^{k,p}_\delta$ norm is equivalent to*

$$\sum_{0 \leq j \leq k} \|\rho^{-\delta} \nabla^j u\|_{0,p;U}.$$

(b) *If $0 \leq \alpha < 1$ and $0 < k + \alpha \leq l + \beta$, $u \in C^{k,\alpha}_\delta(U; E)$ if and only if $\rho^{-\delta} \nabla^j u \in C^{0,\alpha}(U; E \otimes T^j M)$ for $0 \leq j \leq k$, and the $C^{k,\alpha}_\delta$ norm is equivalent to*

$$\sum_{0 \leq j \leq k} \sup_U |\rho^{-\delta} \nabla^j u| + \|\rho^{-\delta} \nabla^k u\|_{0,\alpha;U}$$

PROOF. First consider part (a). By definition, $u \in H^{k,p}_\delta(U; E)$ iff $\nabla^j(\rho^{-\delta} u) \in L^p(U; E \otimes T^j M)$ for $0 \leq j \leq k$. By the product rule and induction, we can write

$$\nabla^j(\rho^{-\delta} u) = \rho^{-\delta} \sum_{\substack{0 \leq i \leq j \\ i+j_1+\cdots+j_p=j}} C(\delta, i, j_1, \ldots, j_p) \nabla^i u \otimes \frac{\nabla^{j_1} \rho}{\rho} \otimes \ldots \otimes \frac{\nabla^{j_p} \rho}{\rho}, \qquad (3.3)$$

where $C(\delta, i, j_1, \ldots, j_p)$ is a constant, equal to 1 when $i = j$. Since, by the preceding lemma, $|\nabla^j \rho|/\rho$ is bounded as long as $j \leq l$, the result follows easily by induction.

For part (b), the case $\delta = 0$ follows by inspecting the coordinate expression for $\nabla^j u$ in Möbius coordinates, recalling that the Christoffel symbols of g are uniformly bounded in $C^{l-1,\beta}$ in these coordinates. The general case follows as above from (3.3) and the fact that $\nabla^j \rho/\rho \in C^{l-j,\beta}(U; T^j M)$. □

It follows from Lemma 2.1 that the norm $\|\cdot\|_{k,0}$ is equivalent to the usual intrinsic C^k norm $\sum_{0 \leq i \leq k} \sup_M |\nabla^i u|$ for $0 \leq k \leq l$. Moreover, if Φ and $\widetilde{\Phi}$ are any Möbius charts whose images intersect, the overlap map $\Phi^{-1} \circ \widetilde{\Phi}$ induces an isomorphism on the space $C^{l,\beta}(U)$ on its domain of definition U, with norm bounded by a uniform constant. Therefore, for $0 \leq k+\alpha \leq l+\beta$, it is immediate that the pullback by any Möbius chart induces isomorphisms on the unweighted Hölder spaces, with norms bounded above and below by constants independent of i. Moreover, if Φ_i is a Möbius chart centered at p_i, $\Phi_i^*\rho = \rho(p_i)y$, which is uniformly bounded above and below on B_2 by constant multiples of $\rho(p_i)$. Therefore the weighted norms scale as follows: for any $r \leq 2$,

$$C^{-1}\rho(p_i)^{-\delta}\|\Phi_i^*u\|_{k,\alpha;B_r} \leq \|u\|_{k,\alpha,\delta;V_r(p_0)} \leq C\rho(p_i)^{-\delta}\|\Phi_i^*u\|_{k,\alpha;B_r}. \qquad (3.4)$$

Similarly, it follows directly from Lemma 2.1 that

$$C^{-1}\rho(p_i)^{-\delta}\|\Phi_i^*u\|_{k,p;B_r} \leq \|u\|_{k,p,\delta;V_r(p_0)} \leq C\rho(p_i)^{-\delta}\|\Phi_i^*u\|_{k,p;B_r}. \qquad (3.5)$$

The next lemma is a strengthening of this.

LEMMA 3.5. *Suppose $\{\Phi_i = \Phi_{p_i}\}$ is a uniformly locally finite cover of M by Möbius charts as in Lemma 2.2. Then we have the following norm equivalences for any r with $1 \leq r \leq 2$:*

$$C^{-1}\sum_i \rho(p_i)^{-\delta}\|\Phi_i^*u\|_{k,p;B_r} \leq \|u\|_{k,p,\delta} \leq C\sum_i \rho(p_i)^{-\delta}\|\Phi_i^*u\|_{k,p;B_r}, \qquad (3.6)$$

$$C^{-1}\sup_i \rho(p_i)^{-\delta}\|\Phi_i^*u\|_{k,\alpha;B_r} \leq \|u\|_{k,\alpha,\delta} \leq C\sup_i \rho(p_i)^{-\delta}\|\Phi_i^*u\|_{k,\alpha;B_r}, \qquad (3.7)$$

PROOF. If Φ_i is a Möbius chart centered at p_i, $\Phi_i^*\rho = \rho(p_i)y$, which is uniformly bounded above and below by constant multiples of $\rho(p_i)$. Thus if $u \in H_\delta^{k,p}(M;E)$, (3.5) yields

$$\sum_i \rho(p_i)^{-\delta}\|\Phi_i^*u\|_{k,p;B_r} \leq C\sum_i \|\Phi_i^*(\rho^{-\delta}u)\|_{k,p;B_r}$$

$$\leq C'\sum_i \|\rho^{-\delta}u\|_{k,p;V_r(p_i)}$$

$$\leq C'N\|\rho^{-\delta}u\|_{k,p}$$

$$= C'N\|u\|_{k,p,\delta},$$

where N is an upper bound on the number of sets $V_2(p_i)$ that can intersect non-trivially. Conversely, if $\sum_i \rho(p_i)^{-\delta}\|\Phi_i^*u\|_{k,p;B_r}$ is finite, then

$$\|u\|_{k,p,\delta} \leq \sum_i \|\rho^{-\delta}u\|_{k,p;V_r(p_i)}$$

$$\leq C\sum_i \|\Phi_i^*(\rho^{-\delta}u)\|_{k,p;B_r}$$

$$\leq C'\sum_i \rho(p_i)^{-\delta}\|\Phi_i^*u\|_{k,p;B_r}.$$

The argument for the Hölder case is similar but simpler, because we do not need the uniform local finiteness in that case. □

The following results follow easily from Lemma 3.5 together with standard facts about Sobolev and Hölder spaces on $B_r \subset \mathbb{H}$, so the proofs are left to the reader (cf. [**29**, **7**]).

LEMMA 3.6. *Let U be any open subset of M, and let E, E_1, E_2 be geometric tensor bundles over M.*

(a) If $1 < p < \infty$, $0 \leq \alpha < 1$, $\delta, \delta' \in \mathbb{R}$, $0 \leq k + \alpha \leq l + \beta$, and $1 \leq k' + \alpha \leq l + \beta$, the pointwise tensor product induces continuous maps
$$C_\delta^{k,\alpha}(U; E_1) \times H_{\delta'}^{k,p}(U; E_2) \to H_{\delta+\delta'}^{k,p}(U; E_1 \otimes E_2),$$
$$C_\delta^{k',\alpha}(U; E_1) \times C_{\delta'}^{k',\alpha}(U; E_2) \to C_{\delta+\delta'}^{k',\alpha}(U; E_1 \otimes E_2).$$

(b) If $1 < p, p' < \infty$, $0 \leq \alpha < 1$, and $0 \leq k + \alpha \leq l + \beta$, we have continuous inclusions:
$$H_\delta^{k,p}(U; E) \hookrightarrow H_{\delta'}^{k,p'}(U; E), \qquad p \geq p', \quad \delta + \frac{n}{p} > \delta' + \frac{n}{p'};$$
$$C_\delta^{k,\alpha}(U; E) \hookrightarrow H_{\delta'}^{k,p'}(U; E), \qquad\qquad \delta > \delta' + \frac{n}{p'}.$$

(c) (SOBOLEV EMBEDDING) If $1 < p < \infty$, $0 < \alpha < 1$, $1 \leq k \leq l$, $k + \alpha \leq l + \beta$, and $\delta \in \mathbb{R}$, we have continuous inclusions
$$H_\delta^{k,p}(U; E) \hookrightarrow C_\delta^{j,\alpha}(U; E), \qquad k - \frac{n+1}{p} \geq j + \alpha,$$
$$H_\delta^{k,p}(U; E) \hookrightarrow H_\delta^{j,p'}(U; E), \qquad k - \frac{n+1}{p} \geq j - \frac{n+1}{p'}.$$

(d) (RELLICH LEMMA) If $1 < p < \infty$, $0 < \alpha < 1$, $0 \leq k \leq l$, and $0 < j + \alpha \leq l + \beta$, then the following inclusions are compact operators:
$$H_\delta^{k,p}(M; E) \hookrightarrow H_{\delta'}^{k',p}(M; E), \qquad k > k', \quad \delta > \delta',$$
$$C_\delta^{j,\alpha}(M; E) \hookrightarrow C_{\delta'}^{j',\alpha}(M; E), \qquad j > j', \quad \delta > \delta'.$$

The following relationships between the Hölder spaces on M and those on $\overline{M}$ will play an important role in Chapter 8.

LEMMA 3.7. *Let E be a geometric tensor bundle of weight r over $\overline{M}$, and suppose $0 < \alpha < 1$, $0 < k + \alpha \leq l + \beta$, and $0 \leq s \leq k + \alpha$. The following inclusions are continuous.*

(a) $C_{(s)}^{k,\alpha}(\overline{M}; E) \hookrightarrow C_{s+r}^{k,\alpha}(M; E)$.
(b) $C_{k+\alpha+r}^{k,\alpha}(M; E) \hookrightarrow C_{(0)}^{k,\alpha}(\overline{M}; E)$.

PROOF. We will prove (a) by induction on k. Suppose $k = 0$, and consider first the case of scalar functions, so that E is the trivial line bundle. By Lemma 3.1(c) there are only two distinct cases: $s = 0$ and $s = \alpha$. For $s = 0$, let Φ_{p_0} be a Möbius chart and let (ρ_0, θ_0) be the background coordinates of p_0. We estimate

$$|\Phi_{p_0}^* u(x, y)| \leq \sup_M |u|$$
$$\leq \|u\|_{C_{(0)}^{0,\alpha}(\overline{M})}. \qquad (3.8)$$
$$|\Phi_{p_0}^* u(x, y) - \Phi_{p_0}^* u(x', y')| = |u(\theta_0 + \rho_0 x, \rho_0 y) - u(\theta_0 + \rho_0 x', \rho_0 y')|$$
$$\leq \|u\|_{C_{(0)}^{0,\alpha}(\overline{M})} |(\rho_0 x, \rho_0 y) - (\rho_0 x', \rho_0 y')|^\alpha$$
$$\leq \|u\|_{C_{(0)}^{0,\alpha}(\overline{M})} \rho_0^\alpha |(x, y) - (x', y')|^\alpha. \qquad (3.9)$$

Since ρ_0^α is uniformly bounded on M, the result follows.

When $s = \alpha$, we need to show that $\Phi_{p_0}^*(\rho^{-\alpha}u)$ is uniformly bounded in $C^{0,\alpha}(B_2)$. The Hölder estimate for u in background coordinates, together with the fact that u vanishes on ∂M, shows that

$$|u(\theta, \rho)| \leq \|u\|_{C^{0,\alpha}_{(0)}(\overline{M})} \rho^\alpha$$

near ∂M, from which the zero-order estimate

$$|\Phi_{p_0}^*(\rho^{-\alpha}u)| \leq C\|u\|_{C^{0,\alpha}_{(0)}(\overline{M})}$$

follows immediately. The Hölder estimate is proved by noting that $\Phi_{p_0}^*(\rho^{-\alpha}u) = y^{-\alpha}\rho_0^{-\alpha}\Phi_{p_0}^*u$; since $y^{-\alpha}$ is uniformly bounded together with all derivatives on B_2, it suffices to show that $\rho_0^{-\alpha}\Phi_{p_0}^*u$ is uniformly bounded in $C^{0,\alpha}(B_2)$. This is proved as follows:

$$\begin{aligned}|\rho_0^{-\alpha}\Phi_{p_0}^*u(x,y) - \rho_0^{-\alpha}\Phi_{p_0}^*u(x',y')| &= \rho_0^{-\alpha}|u(\theta_0 + \rho_0 x, \rho_0 y) - u(\theta_0 + \rho_0 x, \rho_0 y)| \\ &\leq \rho_0^{-\alpha}\|u\|_{C^{0,\alpha}_{(0)}(\overline{M})}|(\rho_0 x, \rho_0 y) - (\rho_0 x, \rho_0 y)|^\alpha \\ &= \|u\|_{C^{0,\alpha}_{(0)}(\overline{M})}|(x,y) - (x',y')|^\alpha.\end{aligned}$$

Now let E be a tensor bundle of type $\binom{p}{q}$ (and thus of weight $r = p - q$). In any background coordinate domain Ω, basis tensors of the form $d\theta^{i_1} \otimes \cdots \otimes d\theta^{i_p} \otimes \partial_{\theta^{j_1}} \otimes \cdots \otimes \partial_{\theta^{j_q}}$ are easily seen to be uniformly bounded in $C^{l,\beta}_r(\Omega; E)$. Since any $u \in C^{0,\alpha}_{(s)}(\overline{M}; E)$ can be written locally as a linear combination of such tensors multiplied by functions in $C^{0,\alpha}_{(s)}(\Omega)$, the result follows from the scalar case together with Lemma 3.6(a).

Now suppose $1 < k + \alpha \leq l + \beta$ and $0 \leq s \leq k + \alpha$, and assume claim (a) is true for $C^{k_0,\alpha}_{(s_0)}(\overline{M}; E)$ when $0 \leq k_0 < k$ and $0 \leq s_0 \leq k_0 + \alpha$. If $u \in C^{k,\alpha}_{(s)}(\overline{M}; E)$, then $|u|_g = \rho^r|u|_{\overline{g}} = O(\rho^{s+r})$ by definition, so it suffices to prove that $\nabla u \in C^{k-1,\alpha}_{s+r}(M; E \otimes T^*M)$ with norm bounded by $\|u\|_{C^{k,\alpha}_{(s)}(\overline{M};E)}$. If $D = \nabla - \overline{\nabla}$ denotes the difference tensor between the Levi-Civita connections of g and $\overline{g}$, a computation shows that D is the 3-tensor whose components are given in any coordinates by

$$D^k_{ij} = -\rho^{-1}(\delta^k_i \partial_j \rho + \delta^k_j \partial_i \rho - \overline{g}^{kl}\overline{g}_{ij}\partial_l \rho). \tag{3.10}$$

Working in background coordinates, we find that $\rho Du \in C^{k,\alpha}_{(s)}(M; E \otimes T^*M)$ since the coefficients of ρD are in $C^{l,\beta}_{(0)}(\overline{M})$. Since $\overline{\nabla}u \in C^{k-1,\alpha}_{(s-1)}(M; E \otimes T^*M)$ by Lemma 3.1(e), we use Lemma 3.1(f) to conclude that

$$\begin{aligned}\rho\nabla u &= \rho\overline{\nabla}u + \rho Du \\ &\in \rho C^{k-1,\alpha}_{(s-1)}(\overline{M}; E \otimes T^*M) + C^{k,\alpha}_{(s)}(\overline{M}; E \otimes T^*M) \\ &\subset C^{k-1,\alpha}_{(s)}(\overline{M}; E \otimes T^*M)\end{aligned}$$

with norm bounded by a multiple of $\|u\|_{C^{k,\alpha}_{(s)}(\overline{M};E)}$. Therefore, since $\rho\nabla u$ is a tensor field of weight $r + 1$, the inductive hypothesis implies that $\rho\nabla u \in C^{k-1,\alpha}_{s+r+1}(M; E \otimes T^*M)$, which implies in turn that $\nabla u \in C^{k-1,\alpha}_{s+r}(M; E \otimes T^*M)$ as desired.

For part (b), we begin with the scalar case, and proceed by induction on k. Let $k = 0$, and suppose $u \in C^{0,\alpha}_\alpha(M)$. Given any Möbius chart $\Phi_{p_0}: B_2 \to V_2(p_0)$, let

(θ_0, ρ_0) be the background coordinates of p_0, and let v be the function $\Phi^*(\rho^{-\alpha}u)$ on B_2, so that
$$u(\theta, \rho) = \rho^\alpha v((\theta - \theta_0)/\rho_0, \rho/\rho_0).$$
The hypothesis means that $v \in C^{0,\alpha}(B_2)$, with norm bounded by $\|u\|_{0,\alpha,\alpha}$. For $(\theta, \rho) \in V_2(p_0)$, we estimate
$$|u(\theta, \rho)| = |\rho^\alpha v((\theta - \theta_0)/\rho_0, \rho/\rho_0)| \le \rho^\alpha \|u\|_{0,\alpha,\alpha}. \tag{3.11}$$
Since ρ is bounded above on M, this shows in particular that $\sup |u|$ is bounded by a multiple of $\|u\|_{0,\alpha,\alpha}$. Similarly, if (θ, ρ) and (θ', ρ') are in $V_2(p_0)$, we have
$$\begin{aligned}|u(\theta, \rho) - u(\theta', \rho')| &= |\rho^\alpha v((\theta - \theta_0)/\rho_0, \rho/\rho_0) - \rho'^\alpha v((\theta' - \theta_0)/\rho_0, \rho'/\rho_0)| \\ &\le \rho^\alpha |v((\theta - \theta_0)/\rho_0, \rho/\rho_0) - v((\theta' - \theta_0)/\rho_0, \rho'/\rho_0)| \\ &\quad + (\rho^\alpha - \rho'^\alpha)|v((\theta' - \theta_0)/\rho_0, \rho'/\rho_0)| \\ &\le C\|u\|_{0,\alpha,\alpha} |(\theta, \rho) - (\theta', \rho')|^\alpha.\end{aligned} \tag{3.12}$$

To complete the $k = 0$ case, it suffices to extend (3.12) to any (θ, ρ) and (θ', ρ') that lie in the same background chart, for then u extends continuously to the boundary as an element of $C^{0,\alpha}_{(0)}(\overline{M})$. Note that there is a real number $\gamma \in (1, 2)$ such that whenever $|(\theta, \rho) - (\theta', \rho')| \le \gamma\rho$, the points (θ, ρ) and (θ', ρ') lie in the image of the same Möbius chart. We estimate as follows:
$$|u(\theta, \rho) - u(\theta', \rho')| \le |u(\theta, \rho) - u(\theta', \rho)| + |u(\theta', \rho) - u(\theta', \rho')|. \tag{3.13}$$
For the first term, the case in which $|\theta - \theta'| \le \gamma\rho$ is taken care of by (3.12). On the other hand, if $|\theta - \theta'| \ge \gamma\rho$, (3.11) gives
$$\begin{aligned}|u(\theta, \rho) - u(\theta', \rho)| &\le |u(\theta, \rho)| + |u(\theta', \rho)| \\ &\le 2\rho^\alpha \|u\|_{0,\alpha,\alpha} \\ &\le 2\gamma^{-\alpha} \|u\|_{0,\alpha,\alpha} |\theta - \theta'|^\alpha \\ &\le C\|u\|_{0,\alpha,\alpha} |(\theta, \rho) - (\theta', \rho')|^\alpha.\end{aligned}$$
To estimate the second term of (3.13), let N be a positive integer such that ρ' lies in the interval $[\gamma^N \rho, \gamma^{N+1} \rho]$. Then, since $(\theta', \gamma^i \rho)$ and $(\theta', \gamma^{i+1} \rho)$ lie in the image of a single Möbius chart, as do $(\theta', \gamma^N \rho)$ and (θ', ρ'), (3.12) gives
$$\begin{aligned}|u(\theta', \rho') - u(\theta', \rho)| &\le |u(\theta', \rho') - u(\theta', \gamma^N \rho)| + \sum_{i=0}^{N-1} |u(\theta', \gamma^{i+1}\rho) - u(\theta', \gamma^i \rho)| \\ &\le C\|u\|_{0,\alpha,\alpha} \left(|\rho' - \gamma^N \rho|^\alpha + \sum_{i=0}^{N-1} |\gamma^{i+1}\rho - \gamma^i \rho|^\alpha\right) \\ &= C\|u\|_{0,\alpha,\alpha} \left(|\rho' - \gamma^N \rho|^\alpha + (\gamma-1)^\alpha \rho^\alpha \sum_{i=0}^{N-1} \gamma^{\alpha i}\right) \\ &= C\|u\|_{0,\alpha,\alpha} \left(|\rho' - \gamma^N \rho|^\alpha + (\gamma-1)^\alpha \rho^\alpha \frac{\gamma^{\alpha N} - 1}{\gamma^\alpha - 1}\right) \\ &\le C'\|u\|_{0,\alpha,\alpha} \left(|\rho' - \gamma^N \rho|^\alpha + |\gamma^N \rho - \rho|^\alpha\right),\end{aligned}$$
where the last inequality follows from the fact that $\gamma^{\alpha N} - 1 \le C(\gamma^N - 1)^\alpha$ when $N \ge 1$. Since both terms in parentheses above are bounded by $|\rho - \rho'|^\alpha$ and thus by $|(\theta, \rho) - (\theta', \rho')|^\alpha$, this completes the argument for the $k = 0$ case. (I am indebted

to Eric Bahuaud for pointing out a gap in an earlier version of this proof, and providing helpful suggestions for fixing it.)

For $k \geq 1$, to show that $u \in C^{k,\alpha}_{(0)}(\overline{M})$, it suffices to show that u is bounded and $Vu \in C^{k-1,\alpha}_{(0)}(\overline{M})$ whenever V is a smooth vector field on $\overline{M}$. It is straightforward to check that any such vector field maps $C^{k,\alpha}_{k+\alpha}(M)$ into $C^{k-1,\alpha}_{k-1+\alpha}(M)$. Thus if $u \in C^{k,\alpha}_{k+\alpha}(M)$, then $Vu \in C^{k-1,\alpha}_{k-1+\alpha}(M) \subset C^{k-1,\alpha}_{(0)}(\overline{M})$ by induction, with norm bounded by $\|u\|_{k,\alpha,k+\alpha}$. Since $|u|$ is obvious uniformly bounded by $\|u\|_{k,\alpha,k+\alpha}$, it follows that $u \in C^{k,\alpha}_{(0)}(\overline{M})$.

Finally, let E be a $\binom{p}{q}$ tensor bundle. If $u \in C^{k,\alpha}_{k+\alpha+r}(M;E)$, to show that $u \in C^{k,\alpha}_{(0)}(\overline{M};E)$ we have to show that the components of u in any background coordinate domain Ω are in $C^{k,\alpha}_{(0)}(\Omega)$. These components are given by complete contractions of tensor products of u with tensors of the form $\partial_{\theta^{i_1}} \otimes \cdots \otimes \partial_{\theta^{i_p}} \otimes d\theta^{j_1} \otimes \cdots \otimes d\theta^{j_q}$, each of which is in $C^{l,\beta}_{-r}(\Omega; T^q_p M)$. By Lemma 3.6(a), these tensor products are in $C^{k,\alpha}_{k+\alpha}(\Omega; E \otimes T^q_p M)$. Since complete contraction clearly maps this space into $C^{k,\alpha}_{k+\alpha}(\Omega)$, the result for tensor fields follows from the scalar case. $\square$

LEMMA 3.8. *Let $\psi \colon \mathbb{R}^+ \to [0,1]$ be a smooth function that is equal to 1 on $[0, \frac{1}{2}]$ and supported in $[0,1)$, and set $\psi_\varepsilon(q) = \psi(\rho(q)/\varepsilon)$ for $q \in M$. If $0 \leq \alpha < 1$ and $0 \leq k + \alpha \leq l + \beta$, then $\psi_\varepsilon \in C^{k,\alpha}(M)$, with norm bounded independently of ε.*

PROOF. Working directly with the definition of $C^{k,\alpha}(M)$, for any Möbius coordinate chart Φ_{p_0}, we have to show that $\Phi^*_{p_0}\psi_\varepsilon(x,y) = \psi(\rho(p_0)y/\varepsilon)$ is uniformly bounded in $C^{k,\alpha}(B_2)$. Since $e^{-2} < y < e^2$ on B_2, $\Phi^*_{p_0}\psi_\varepsilon$ will be identically equal to 0 or 1 on B unless $\frac{1}{2}\varepsilon e^{-2} < \rho(p_0) < \varepsilon e^2$. Under these restrictions, it is easy to verify that $\psi(\rho(p_0)y/\varepsilon)$ is uniformly bounded in $C^{k,\alpha}(B_2)$. $\square$

LEMMA 3.9. *If $1 < p < \infty$, $\delta \in \mathbb{R}$, and $0 \leq k \leq l$, the set of compactly supported smooth tensor fields is dense in $H^{k,p}_\delta(M;E)$.*

PROOF. Suppose $u \in H^{k,p}_\delta(M;E)$. First we show that u can be approximated in the $H^{k,p}_\delta$ norm by compactly supported elements of $H^{k,p}_\delta(M;E)$.

Let ψ_ε be as in the preceding lemma. We will show that $(1 - \psi_\varepsilon)u \to u$ in $H^{k,p}_\delta$ as $\varepsilon \to 0$, which by Lemma 3.4 is the same as $\nabla^j(\psi_\varepsilon u) \to 0$ in $H^{0,p}_\delta$ whenever $0 \leq j \leq k$. By the product rule,

$$\nabla^j(\psi_\varepsilon u) = \sum_{i=0}^{j} C_i \nabla^{j-i}\psi_\varepsilon \otimes \nabla^i u.$$

Since $|\nabla^{j-i}\psi_\varepsilon|_g$ is bounded and supported where $\rho \leq 2\varepsilon$, we have

$$\|\nabla^{j-i}\psi_\varepsilon \otimes \nabla^i u\|^p_{0,p,\delta} \leq C \int_{\{\rho \leq 2\varepsilon\}} |\rho^{-\delta}\nabla^i u|^p_g \, dV_g.$$

By Lemma 3.4, $|\rho^{-\delta}\nabla^i u|^p_g$ is integrable, so the integral on the right-hand side above goes to zero as $\varepsilon \to 0$ by the dominated convergence theorem.

Next we must check that if $u \in H^{k,p}_\delta(M;E)$ is compactly supported in M, it can be approximated in the $H^{k,p}_\delta$ norm by smooth, compactly supported tensor fields. The classical argument involving convolution with an approximate identity shows that u can be approximated in the standard Sobolev $H^{k,p}$ norm on a slightly larger compact set by tensor fields in $C_c^\infty(M;E)$. However, on any fixed compact subset of M, it is easy to see that the $H^{k,p}_\delta$ norm is equivalent to the $H^{k,p}$ norm, thus completing the proof. □

CHAPTER 4

Elliptic Operators

In this chapter, we collect some basic facts regarding elliptic operators on asymptotically hyperbolic manifolds. Throughout this chapter we assume (M,g) is a connected asymptotically hyperbolic $(n+1)$-manifold of class $C^{l,\beta}$ for some $l \geq 2$ and $0 \leq \beta < 1$, and ρ is a fixed smooth defining function.

Let E and F be geometric tensor bundles over M. We will say a linear partial differential operator $P\colon C^\infty(M; E) \to C^\infty(M; F)$ is a *geometric operator of order* m if the components of Pu in any coordinate frame are given by linear functions of the components of u and their partial derivatives, whose coefficients are universal polynomials in the components of the metric g, their partial derivatives, and the function $(\det g_{ij})^{-1}$ (or $(\det g_{ij})^{-1/2}$ if M is oriented), such that the coefficient of any jth derivative of u involves at most the first $m-j$ derivatives of the metric. A geometric operator is, in particular, an example of a *regular natural differential operator* in the terminology introduced by P. Stredder [55]. We note that such an operator is automatically invariant under (orientation-preserving) isometries. By the results of [55], P is geometric of order m if and only if Pu can be written as a sum of contractions of tensors of the form

$$\nabla^j u \otimes \nabla^{k_1} Rm \otimes \cdots \otimes \nabla^{k_l} Rm \otimes$$
$$\underbrace{g \otimes \cdots \otimes g}_{p \text{ factors}} \otimes \underbrace{g^{-1} \otimes \cdots \otimes g^{-1}}_{q \text{ factors}} \otimes \underbrace{dV_g \otimes \cdots \otimes dV_g}_{s \text{ factors}}, \quad (4.1)$$

(possibly after reordering indices), with $0 \leq j \leq m$ and $0 \leq k_i \leq m-j-2$, and with $s = 0$ unless M is oriented. Here Rm is the (covariant) Riemann curvature tensor of g, and dV_g is its Riemannian volume form.

If E and F are tensor bundles over $\overline{M}$, a differential operator $P\colon C^\infty(M; E) \to C^\infty(M; F)$ is said to be *uniformly degenerate* if it can be written in background coordinates as a system of operators that are polynomials in $\rho\partial/\partial\theta^i$ with coefficients that are at least continuous up to the boundary.

LEMMA 4.1. *Let (M,g) be an asymptotically hyperbolic $(n+1)$-manifold of class $C^{l,\beta}$, $0 \leq \beta < 1$. Suppose E is a geometric tensor bundle over M, and $P\colon C^\infty(M;E) \to C^\infty(M;E)$ is a geometric operator of order $m \leq l$. Then P is uniformly degenerate.*

PROOF. As noted above, for any section u of E, Pu can be written as a sum of contractions of terms like (4.1). If u is covariant of degree r_1 and contravariant of degree r_2, then this tensor product has $r_2 + 2q$ upper indices and $r_1 + j + 2p + (n+1)s + 4l + \sum_i k_i$ lower indices. (The $4l$ lower indices are the undifferentiated indices of the l copies of Rm.) Because we are assuming that Pu is the same type of tensor as u, $2q$ of the upper indices must be contracted against $j + 2p + (n+1)s + 4l + \sum_i k_i$

of the lower indices, so in particular we must have

$$2q = j + 2p + (n+1)s + 4l + \sum_i k_i. \tag{4.2}$$

It is obvious that tensoring with $\overline{g} = \rho^2 g$, $\overline{g}^{-1} = \rho^{-2} g^{-1}$, and $dV_{\overline{g}} = \rho^{n+1} dV_g$ are all uniformly degenerate operators. Using formula (3.10) for the components of the difference tensor $D = \nabla - \overline{\nabla}$, we see that the components of ρD_{ij}^k in background coordinates are $C^{l,\beta}$ up to the boundary. It follows that $(\rho\nabla)^j = (\rho\overline{\nabla} + \rho D)^j$ is a uniformly degenerate operator for $0 \leq j \leq l$. Since $[\nabla, \rho]u = u \otimes d\rho$ is also uniformly degenerate, it follows by induction that $\rho^j \nabla^j$ is uniformly degenerate as well.

A straightforward computation (cf. [**40, 29**]) shows that the components of Rm are given by

$$R_{ijkl} = -|d\rho|_{\overline{g}}^2 (g_{ik}g_{jl} - g_{il}g_{jk}) + \rho^{-3} p_1(\overline{g}, \overline{g}^{-1}, \partial \overline{g}) + \rho^{-2} p_2(\overline{g}, \overline{g}^{-1}, \partial \overline{g}, \partial^2 \overline{g}), \tag{4.3}$$

where p_1 and p_2 are universal polynomials, so $\rho^4 Rm \in C^{l-2,\beta}_{(0)}(\overline{M}, T^4\overline{M})$. Moreover, an easy induction shows that for $k \leq l - 2$, $\rho^{4+k} \nabla^k Rm \in C^{l-2-k,\beta}_{(0)}(\overline{M}, T^4\overline{M})$. It follows that tensoring with $\rho^{4+k} \nabla^k Rm$ is a uniformly degenerate operator. By virtue of (4.2), therefore, we can rewrite (4.1) in the following manifestly uniformly degenerate form:

$$\rho^j \nabla^j u \otimes \rho^{4+k_1} \nabla^{k_1} Rm \otimes \cdots \otimes \rho^{4+k_l} \nabla^{k_l} Rm \otimes \rho^2 g \otimes \cdots \otimes \rho^2 g \otimes$$
$$\rho^{-2} g^{-1} \otimes \cdots \otimes \rho^{-2} g^{-1} \otimes \rho^{n+1} dV_g \otimes \cdots \otimes \rho^{n+1} dV_g. \tag{4.4}$$

Since contraction of a lower index against an upper one is also uniformly degenerate, the result follows. □

If P is a uniformly degenerate operator, for each $s \in \mathbb{C}$ we define the *indicial map* of P to be the bundle map $I_s(P) \colon E|_{\partial M} \to E|_{\partial M}$ defined by

$$I_s(P)(\overline{u}) = (\rho^{-s} P(\rho^s \overline{u}))|_{\partial M},$$

where ρ is any smooth defining function, and $\overline{u}$ is extended arbitrarily to a C^m section of E in a neighborhood of ∂M. It is easy to check that the indicial map of any uniformly degenerate operator is a continuous bundle map, which is independent of the extension or the choice of defining function. For geometric operators, we can say more.

LEMMA 4.2. *Suppose (M,g) is an asymptotically hyperbolic $(n+1)$-manifold of class $C^{l,\beta}$, and $P \colon C^\infty(M;E) \to C^\infty(M;E)$ is a geometric operator of order $m \leq l$. Then for each $s \in \mathbb{C}$, $I_s(P) \colon E|_{\partial M} \to E|_{\partial M}$ is a $C^{l,\beta}$ bundle map.*

PROOF. It is an immediate consequence of the definition of the indicial map that if P_1 and P_2 are uniformly degenerate operators with sufficiently smooth coefficients, then

$$I_s(P_1 \circ P_2) = I_s(P_1) \circ I_s(P_2),$$
$$I_s(P_1 + P_2) = I_s(P_1) + I_s(P_2).$$

Since a sum or composition of two $C^{l,\beta}$ bundle maps is again of class $C^{l,\beta}$, it suffices to display P as a sum of compositions of uniformly degenerate operators with $C^{l,\beta}$ indicial maps.

Writing P as a sum of contractions of terms of the form (4.4), we see that it suffices to show that each of the following uniformly degenerate operators has $C^{l,\beta}$ indicial map:

(a) $u \mapsto \rho^j \nabla^j u$;
(b) $u \mapsto u \otimes \rho^{4+k} \nabla^k Rm$.
(c) $u \mapsto u \otimes \rho g$;
(d) $u \mapsto u \otimes \rho^{-1} g^{-1}$;
(e) $u \mapsto u \otimes \rho^{n+1} dV_g$.

The last three operators above are themselves $C^{l,\beta}$ bundle maps over $\overline{M}$, whose indicial maps are just their restrictions to ∂M, so there is nothing to prove in those cases. For (a), observe that $\rho^j \nabla^j u$ can be written as a sum of compositions of the operators $[\rho, \nabla]$ and $\rho \nabla$. Since the commutator $[\rho, \nabla] u = -u \otimes d\rho$ is a smooth bundle map and therefore has smooth indicial map, we need only consider the operator $\rho \nabla$. Let $D = \nabla - \overline{\nabla}$ be the difference tensor as in the preceding proof, and observe that for any $\overline{u} \in C^1_{(0)}(\overline{M}; E)$,

$$\rho^{-s}(\rho \nabla(\rho^s \overline{u})) = \rho^{-s}(\rho \overline{\nabla}(\rho^s \overline{u}) + \rho D(\rho^s \overline{u}))$$
$$= s\overline{u} \otimes d\rho + \rho D(\overline{u}) + O(\rho).$$

As noted above, $\overline{u} \mapsto \overline{u} \otimes d\rho$ is a smooth bundle map; and it follows from (3.10) that ρD is a $C^{l,\beta}$ bundle map. Therefore the indicial map of $\rho \nabla$ is $C^{l,\beta}$ as claimed.

To analyze the remaining operator, $u \mapsto u \otimes \rho^{4+k} \nabla^k Rm$, let K be the 4-tensor field on M whose components are

$$K_{ijkl} = -(g_{ik} g_{jl} - g_{il} g_{jk}),$$

and let $\overline{K}$ be

$$\overline{K}_{ijkl} = -(\overline{g}_{ik} \overline{g}_{jl} - \overline{g}_{il} \overline{g}_{jk}) = \rho^4 K_{ijkl}.$$

Because the assumption that g is asymptotically hyperbolic means that $|d\rho|_{\overline{g}} = 1$ on ∂M, we can write $|d\rho|_{\overline{g}} = 1 + \rho v$, where $v \in C^{l-1,\beta}_{(0)}(\overline{M})$. Using (4.3), therefore, we can write

$$Rm = K + \rho^{-3}\overline{V} = \rho^{-4}\overline{K} + \rho^{-3}\overline{V}, \qquad (4.5)$$

where $\overline{V} \in C^{l-2,\beta}_{(0)}(\overline{M}, T^4\overline{M})$. It follows that $I_s(u \mapsto u \otimes \rho^4 Rm)$ is just tensoring with $\overline{K}|_{\partial M}$, which is a $C^{l,\beta}$ bundle map. On the other hand, Lemma 3.7(a) shows that

$$Rm - K = \rho^{-3}\overline{V} \in \rho^{-3} C^{l-2,\beta}_{(0)}(\overline{M}; T^4\overline{M})$$
$$\subset \rho^{-3} C^{l-2,\beta}_4(M; T^4 M) = C^{l-2,\beta}_1(M; T^4 M).$$

Because K is parallel, $\nabla^k Rm = \nabla^k (Rm - K) \in C^{l-2-k,\beta}_1(M; T^4 M)$. In particular, this implies that

$$|\rho^{4+k} \nabla^k Rm|_{\overline{g}} = |\nabla^k Rm|_g = O(\rho), \qquad (4.6)$$

so $\rho^{4+k} \nabla^k Rm$ has vanishing indicial map for $k > 0$. $\square$

A complex number s is called a *characteristic exponent* of P at $\widehat{p} \in \partial M$ if $I_s(P)$ is singular at $\widehat{p}$. Its *multiplicity* is the dimension of the kernel of $I_s(P) \colon E_{\widehat{p}} \to E_{\widehat{p}}$. If s is a characteristic exponent of P somewhere on ∂M, we just say it is a characteristic exponent of P.

Let $(\mathbb{B}, \breve{g})$ be the unit ball model of hyperbolic space as described in Chapter 2, let $\breve{E}$ denote the tensor bundle over $\mathbb{B}$ that corresponds to E (i.e., associated to the same representation of $\mathrm{O}(n+1)$ or $\mathrm{SO}(n+1)$), and let $\check P\colon C^\infty(\mathbb{B};\check E) \to C^\infty(\mathbb{B};\check E)$ be the geometric operator on $\check E$ whose coordinate formula is the same as that of P. Because $\check P$ is invariant under (orientation-preserving) isometries of $\mathbb{B}$, and Euclidean rotations in the unit ball model of hyperbolic space are isometries of $\check g$ that act transitively on the sphere $\mathbb{S}^n = \partial\mathbb{B}$, it follows that the characteristic exponents of $\check P$ and their multiplicities are constant on $\partial\mathbb{B}$. The next lemma shows that the analogous statement holds for P as well.

LEMMA 4.3. *Let (M,g) be a connected asymptotically hyperbolic $(n+1)$-manifold of class $C^{l,\beta}$, and let $P\colon C^\infty(M;E) \to C^\infty(M;E)$ be a geometric operator of order $m \le l$. The characteristic exponents of P and their multiplicities are constant on ∂M, and are the same as those of $\check P$.*

PROOF. Let $\widehat p \in \partial M$ be arbitrary, and let (θ,ρ) be any background coordinates defined on a neighborhood of $\widehat p$ in $\overline M$. By an affine change of background coordinates, we may arrange that $\widehat p = (0,0)$ and $\overline g_{ij} = \delta_{ij}$ at $\widehat p$. Let $\Phi\colon V \to M$ be the Möbius chart given by $(\theta,\rho) = \Phi(x,y) = (x,y)$ in terms of these background coordinates, defined on some neighborhood V of $(0,0)$ in $\overline{\mathbb{H}}$. Let $\widetilde g$ be the Riemannian metric Φ^*g on $V \cap \mathbb{H}$, let $\widetilde P = \Phi^* \circ P \circ \Phi^{-1*}$, and let $I_s(\widetilde P)(\overline u) = y^{-s}\widetilde P(y^s \overline u)$ be the indicial map of $\widetilde P$. Then $I_s(\widetilde P) = \Phi^* \circ I_s(P) \circ \Phi^{-1*}$, so the characteristic exponents and multiplicities of $\widetilde P$ at $(0,0) \in \partial\mathbb{H}$ are the same as those of P at $\Phi(x,0)$. Thus it suffices to show that $I_s(\widetilde P) = I_s(\check P)$ at $(0,0)$.

For convenience, here is a summary of the several metrics being considered in this proof:

g (the given asymptotically hyperbolic metric on M),

$\overline g = \rho^2 g$ (on $\overline M$),

$\widetilde g = \Phi^* g$ (on an open subset of $\mathbb{H}$),

$\check g$ (the hyperbolic metric on $\mathbb{H}$).

Arguing as in the proof of the preceding lemma, it suffices to show that each of the following operators has vanishing indicial map at $(0,0)$:

(a) $u \mapsto \rho \widetilde\nabla u - \rho \check\nabla u$,
(b) $u \mapsto u \otimes (\rho^2 \widetilde g - \rho^2 \check g)$,
(c) $u \mapsto u \otimes (\rho^{-2}\widetilde g^{-1} - \rho^{-2}\check g^{-1})$,
(d) $u \mapsto u \otimes (\rho^{n+1} dV_{\widetilde g} - \rho^{n+1} dV_{\check g})$,
(e) $u \mapsto u \otimes (\rho^{4+k} \widetilde\nabla^k \widetilde{Rm} - \rho^{4+k}\check\nabla^k \check{Rm})$.

If we let $E = \widetilde\nabla - \check\nabla$ be the difference tensor between the two connections, a computation based on (3.10) yields

$$\rho E^i_{jk} = +\partial_j \rho(\overline g^{il}\overline g_{kl} - \delta^{il}\delta_{kl}) + O(\rho).$$

Because $\overline g_{ij} = \delta_{ij}$ at $\widehat p$, it follows that $\rho\widetilde\nabla - \rho\check\nabla$ is a zero-order operator that vanishes at $(0,0)$.

The next three operators, tensoring with (b)–(d), are obviously zero-order operators vanishing at $(0,0)$, so their indicial maps have the same property. To complete the proof, we need to show that the indicial map of (e) vanishes at $(0,0)$. This is obvious from the argument in the preceding proof when $k > 0$, because both $\rho^{4+k}\widetilde\nabla^k\widetilde{Rm}$ and $\rho^{4+k}\check\nabla^k\check{Rm}$ individually have vanishing indicial maps. On the other hand, the preceding proof showed that the restriction of $\rho^4 \widetilde{Rm}$ to ∂M is equal to $\overline{K}$, which in turn is equal to $\rho^4 \check{Rm}$ at $(0,0)$. This completes the proof. □

The next observation we need to make is that if P is a formally self-adjoint geometric operator acting on a geometric tensor bundle of weight r, its set of characteristic exponents is symmetric about the line $\operatorname{Re} s = n/2 - r$. This will follow easily from the next lemma.

PROPOSITION 4.4. *Let (M,g) be an asymptotically hyperbolic $(n+1)$-manifold of class $C^{l,\beta}$. Suppose that $m \leq l$ and $P \colon C^\infty(M;E) \to C^\infty(M;E)$ is an mth order geometric operator acting on sections of a geometric tensor bundle E of weight r, and let P^* denote its formal adjoint. Then*

$$I_s(P^*) = I_{n-2r-\overline{s}}(P)^*.$$

PROOF. Choose any point $\widehat{p} \in \partial M$, and let (θ, ρ) be background coordinates in a neighborhood Ω of $\widehat{p}$ in $\overline{M}$. Then we can write P locally as

$$Pu(\theta, \rho) = \sum_{\substack{0 \leq k \leq m \\ 1 \leq j_i, \ldots, j_k \leq n+1}} a^{j_1 \ldots j_k}(\theta, \rho)(\rho \partial_{j_1}) \cdots (\rho \partial_{j_k}) u(\theta, \rho),$$

where each coefficient $a^{j_1 \ldots j_k}(\theta, \rho)$ is a matrix-valued $C^{l-m+k,\beta}$ function (the coordinate representation of a bundle map from E to itself). Suppose now that $\overline{u}$ is a section of E that has a C^m extension to $\overline{M}$. Then $\rho \partial_i(\rho^s \overline{u}) = O(\rho^{s+1})$ if $i \neq n+1$, so $I_s(P)$ is given locally by

$$I_s(P)\overline{u} = \lim_{\rho \to 0} \sum_{\substack{0 \leq k \leq m \\ j_i = \cdots = j_k = n+1}} \rho^{-s} a^{j_1 \ldots j_k}(\theta, \rho)(\rho \partial_{n+1}) \cdots (\rho \partial_{n+1})(\rho^s \overline{u}(\theta, \rho))$$

$$= \sum_{\substack{0 \leq k \leq m \\ j_i = \cdots = j_k = n+1}} s^k a^{j_1 \ldots j_k}(\theta, 0) \overline{u}(\theta, 0).$$

To compute $I_s(P^*)$, we first compute the formal adjoint of a monomial of the form $\rho \partial_i$ as follows. Let $\overline{g} = \rho^2 g$, which is a $C^{l,\beta}$ metric on $\overline{M}$. The inner products on E defined by g and $\overline{g}$ are related by $\langle u, v \rangle_g = \rho^{2r} \langle u, v \rangle_{\overline{g}}$, and the volume elements by $dV_g = \rho^{-n-1} dV_{\overline{g}}$. If u, v are smooth sections of E compactly supported in $\Omega \cap M$, then

$$\int_M \langle u, \rho \partial_i v \rangle_g dV_g = \int_M \rho^{2r-n-1} \langle u, \rho \partial_i v \rangle_{\overline{g}} (\det \overline{g})^{1/2} d\theta^1 \ldots d\theta^{n+1}$$

$$= -\int_M \langle \partial_i(\rho^{2r-n} u b(\overline{g})), v \rangle_{\overline{g}} (\det \overline{g})^{1/2} d\theta^1 \ldots d\theta^{n+1}$$

$$= -\int_M \langle \rho^{n+1-2r} \partial_i(\rho^{2r-n} u b(\overline{g})), v \rangle_g dV_g,$$

where $b(\overline{g})$ is a constant-coefficient polynomial in the components of $\overline{g}$, $\overline{g}^{-1}$, and $(\det \overline{g})^{1/2}$. From this, it follows easily that

$$(\rho \partial_i)^* = -\rho \partial_i + (n - 2r) \delta_i^{n+1} - \rho b_i,$$

where $b_i = \partial_i b(\overline{g})$ is $C^{l-1,\beta}$ up to ∂M. Therefore,

$$P^* u(\theta, \rho) = \sum_{\substack{0 \le k \le m \\ 1 \le j_i, \ldots, j_k \le n+1}} (-\rho \partial_{j_k} + (n-2r)\delta^{n+1}_{j_k} - \rho b_{j_k}) \cdots$$
$$(-\rho \partial_{j_1} + (n-2r)\delta^{n+1}_{j_1} - \rho b_{j_1})(a^{j_1 \cdots j_k}(\theta, \rho)^* u(\theta, \rho)),$$

and we conclude that

$$I_s(P^*)\overline{u} = \lim_{\rho \to 0} \sum_{\substack{0 \le k \le m \\ j_i = \cdots = j_k = n+1}} \rho^{-s}(-\rho \partial_{n+1} + (n-2r) - \rho b_{n+1}) \cdots$$
$$(-\rho \partial_{n+1} + (n-2r) - \rho b_{n+1})(a^{j_1 \cdots j_k}(\theta, \rho)^*(\rho^s \overline{u}(\theta, \rho)))$$
$$= \sum_{\substack{0 \le k \le m \\ j_i = \cdots = j_k = n+1}} (n - 2r - s)^k a^{j_1 \cdots j_k}(\theta, 0)^* \overline{u}(\theta, 0)$$
$$= I_{n-2r-\overline{s}}(P)^* \overline{u},$$

which was to be proved. □

COROLLARY 4.5. *If P is a formally self-adjoint geometric operator of order $m \le l$, then the set of characteristic exponents of P is symmetric about the line $\operatorname{Re} s = n/2 - r$.*

PROOF. The preceding proposition shows that $I_s(P) = I_s(P^*) = I_{n-2r-\overline{s}}(P)^*$. Thus if s is a characteristic exponent of P, then $I_{n-2r-\overline{s}}(P)^*$, and hence also $I_{n-2r-\overline{s}}(P)$, is singular, which means that $s' = n - 2r - \overline{s}$ is also a characteristic exponent. Since $\operatorname{Im} s = \operatorname{Im} s'$ and $\frac{1}{2}(\operatorname{Re} s + \operatorname{Re} s') = n/2 - r$, the result follows. □

This corollary shows that a geometric self-adjoint operator P must have at least one characteristic exponent whose real part is greater than or equal to $n/2 - r$. We define the *indicial radius* of P to be the smallest nonnegative number R such that P has a characteristic exponent whose real part is $n/2 - r + R$. (This is well-defined because the set of characteristic exponents is finite by Lemma 4.3.)

Next we investigate the mapping properties of geometric operators on our weighted spaces.

LEMMA 4.6. *Let $P \colon C^\infty(M; E) \to C^\infty(M; F)$ be a geometric operator of order m.*

(a) *If $\delta \in \mathbb{R}$, $1 < p < \infty$, and $m \le k \le l$, then P extends naturally to a bounded mapping*
$$P \colon H^{k,p}_\delta(M; E) \to H^{k-m,p}_\delta(M; F).$$

(b) *If $\delta \in \mathbb{R}$, $0 \le \alpha < 1$, and $m \le k + \alpha \le l + \beta$, then P extends naturally to a bounded mapping*
$$P \colon C^{k,\alpha}_\delta(M; E) \to C^{k-m,\alpha}_\delta(M; F).$$

PROOF. Let $\{\Phi_i\}$ be a uniformly locally finite covering of M by Möbius charts as in Lemma 2.2. Let $\breve{E}$ be the bundle over hyperbolic space associated to the same $O(n+1)$ or $SO(n+1)$ representation as E, and for each i, let $P_i \colon C^\infty(B_2; E) \to C^\infty(B_2; E)$ be the operator defined by $P_i = \Phi_i^* \circ P \circ (\Phi_i^{-1})^*$. Since the metric $g_i = \Phi_i^* g$ is uniformly $C^{l,\beta}$ equivalent to $\breve{g}$ in Möbius coordinates by Lemma (2.1), it follows that the coefficients of the jth derivatives of u that appear in $P_i u$ are uniformly bounded in $C^{l-m+j,\beta}(B_2)$. Therefore, using Lemmas 3.5 and 3.6(a), we have

$$\|Pu\|_{k,p,\delta} \leq C \sum_i \rho(p_i)^{-\delta} \|\Phi_i^* Pu\|_{k,p;B_2}$$
$$= C \sum_i \rho(p_i)^{-\delta} \|P_i(\Phi_i^* u)\|_{k,p;B_2}$$
$$\leq C' \sum_i \rho(p_i)^{-\delta} \|\Phi_i^* u\|_{k-m,p;B_2}$$
$$\leq C'' \|u\|_{k-m,p,\delta},$$

with an analogous estimate in the Hölder case. □

Recall from Chapter 3 that when $p^* = p/(p-1)$, $H^{0,p^*}_{-\delta}(M;E)$ is naturally dual to $H^{0,p}_{\delta}(M;E)$ under the standard L^2 pairing. The next lemma shows that P is symmetric with respect to this pairing.

LEMMA 4.7. *Suppose P satisfies the hypotheses of Theorem C. If $1 < p < \infty$, $p^* = p/(p-1)$, and $\delta \in \mathbb{R}$, then $(Pu, v) = (u, Pv)$ for all $u \in H^{0,p}_{\delta}(M;E)$, $v \in H^{0,p^*}_{-\delta}(M;E)$.*

PROOF. This is true for $u, v \in C^\infty_c(M;E)$ by the fact that P is formally self-adjoint. The general result follows by density. □

The following lemma is a standard application of rescaling techniques and classical interior elliptic regularity (cf. [**29**, Prop. 3.4], [**7**, Lemma 2.4], and [**8**, Lemma 4.1.1]).

LEMMA 4.8. *Let $P\colon C^\infty(M;E) \to C^\infty(M;F)$ be a geometric elliptic operator of order m.*

(a) *Suppose $\delta \in \mathbb{R}$, $1 < p < \infty$, and $m \leq k \leq l$. If $u \in H^{0,p}_\delta(M;E)$ is such that $Pu \in H^{k-m,p}_\delta(M;F)$, then $u \in H^{k,p}_\delta(M;E)$ and*

$$\|u\|_{k,p,\delta} \leq C(\|Pu\|_{k-m,p,\delta} + \|u\|_{0,p,\delta}). \tag{4.7}$$

(b) *Suppose $\delta \in \mathbb{R}$, $0 < \alpha < 1$, and $m < k + \alpha \leq l + \beta$. If $u \in C^{0,0}_\delta(M;E)$ is such that $Pu \in C^{k-m,\alpha}_\delta(M;F)$, then $u \in C^{k,\alpha}_\delta(M;E)$ and*

$$\|u\|_{k,\alpha,\delta} \leq C(\|Pu\|_{k-m,\alpha,\delta} + \|u\|_{0,0,\delta}). \tag{4.8}$$

PROOF. Under the hypotheses of case (a), u is locally in $H^{k,p}$ by interior elliptic regularity, so only the estimate (4.7) needs to be proved. Let $\{\Phi_i\}$ be a uniformly locally finite covering of M by Möbius charts as in Lemma 2.2, and let $P_i = \Phi_i^* \circ P \circ (\Phi_i^{-1})^*$ as in the proof of Lemma 4.6. Since the coefficients of the highest-order terms in P_i are constant-coefficient polynomials in $g_i = \Phi_i^* g$ and $(\det g_i)^{-1/2}$, and g_i is uniformly equivalent to the hyperbolic metric by Lemma 2.1, P_i is uniformly elliptic on B_2. Moreover, since the coefficients of P_i are uniformly bounded in $C^{k-m,\alpha}(B_2)$ by the same lemma, we have the following standard local elliptic estimate [**48**, **1**]:

$$\|u\|_{k,p;B_1} \leq C(\|P_i u\|_{k-m,p;B_2} + \|u\|_{0,p;B_2}),$$

where the constant C depends on P, k, p, and δ, but is independent of u and i. Thus, using Lemma 3.5 again, we have

$$\|u\|_{k,p,\delta} \leq C \sum_i \rho(p_i)^{-\delta} \|\Phi_i^* u\|_{k,p;B_1}$$

$$\leq C' \sum_i \rho(p_i)^{-\delta} (\|P_i(\Phi_i^* u)\|_{k-m,p;B_2} + \|\Phi_i^* u\|_{0,p;B_2})$$

$$= C' \sum_i \rho(p_i)^{-\delta} \|\Phi_i^*(Pu)\|_{k-m,p;B_2} + C' \sum_i \rho(p_i)^{-\delta} \|\Phi_i^* u\|_{0,p;B_2}$$

$$\leq C''(\|Pu\|_{k-m,p,\delta} + \|u\|_{0,p,\delta}).$$

The argument for case (b) is similar, using interior Hölder estimates [26]. □

LEMMA 4.9. *If P satisfies the hypotheses of Theorem C, then P is self-adjoint as an unbounded operator on $L^2(M; E)$.*

PROOF. Because of the density of $C_c^\infty(M; E)$ in $H^{m,2}(M; E)$, P is densely defined, and Lemma 4.8 shows that its domain is exactly $H^{m,2}(M; E)$. Clearly the domain of its Hilbert space adjoint contains $H^{m,2}(M; E)$. On the other hand, if v is in the domain of the adjoint, then there exists $w \in L^2(M; E)$ such that $(v, Pu) = (w, u)$ for all $u \in L^2(M; E)$. This means in particular that $Pv = w$ as distributions, which by Lemma 4.8 implies that $v \in H^{m,2}(M; E)$. Thus the domain of the adjoint is equal to the domain of P. □

Next we present some elementary preliminary results about Fredholm operators on these weighted spaces. Our first lemma reduces the problem of proving L^p Fredholm theorems to estimates near the boundary. A partial differential operator P acting on sections of a vector bundle over a connected manifold M is said to have the *weak unique continuation property* if any solution to $Pu = 0$ that vanishes on an open set must vanish on all of M. It has been shown recently by G. Nakamura, G. Uhlmann, and J.-N. Wang [47] that every strongly elliptic operator has this property. We say an operator is *semi-Fredholm* if it has finite-dimensional kernel and closed range.

LEMMA 4.10. *Let $P\colon C^\infty(M; E) \to C^\infty(M; E)$ be a formally self-adjoint geometric elliptic operator of order $m \leq l$. Suppose $1 < p < \infty$ and $\delta \in \mathbb{R}$. Then $P\colon H_\delta^{k,p}(M; E) \to H_\delta^{k-m,p}(M; E)$ is semi-Fredholm for $m \leq k \leq l$ if and only if there exist a compact subset $K \subset M$ and a constant $c > 0$ such that the following estimate holds for all $u \in C_c^\infty(M \smallsetminus K; E)$:*

$$c\|u\|_{0,p,\delta} \leq \|Pu\|_{0,p,\delta}. \tag{4.9}$$

If P is semi-Fredholm and has the weak unique continuation property, then (4.9) holds for every compact subset $K \subset M$ with nonempty interior. If both (4.9) and

$$c\|u\|_{0,p^*,-\delta} \leq \|Pu\|_{0,p^*,-\delta} \tag{4.10}$$

hold for $u \in C_c^\infty(M \smallsetminus K; E)$, where $p^* = p/(p-1)$, then P is Fredholm.

PROOF. The argument in the forward direction is fairly standard; see, for example, [12, Thm. 1.10] and [7, Prop. 2.6]. Let U and V be precompact open subsets of M such that $K \subset U \subset \overline{U} \subset V$, and let $\psi \in C^\infty(M)$ be a smooth bump function that is equal to 1 on K and supported in $\overline{U}$. It is easy to check that multiplication by ψ is a bounded map from $H_\delta^{k,p}(M; E)$ to itself for any k, p, δ. On the compact set $\overline{V}$, the $H^{k,p}_\delta$ norm is uniformly equivalent to the standard $H^{k,p}$ norm, and P is uniformly elliptic.

First observe that (4.9) extends to all $u \in H^{k,p}_\delta(M;E)$ by continuity. For any $u \in H^{k,p}_\delta(M;E)$, we can write $u = u_\infty + u_0$, where $u_0 := \psi u$ is supported in $\overline{U}$ and $u_\infty := (1-\psi)u$ is supported in $M \smallsetminus K$. We estimate as follows:

$$\|u\|_{k,p,\delta} \leq \|u_\infty\|_{k,p,\delta} + \|u_0\|_{k,p,\delta}.$$

For the first term, (4.7) gives

$$\begin{aligned}\|u_\infty\|_{k,p,\delta} &\leq C_1(\|Pu_\infty\|_{k-m,p,\delta} + \|u_\infty\|_{0,p,\delta}) \\ &\leq C_2(\|Pu_\infty\|_{k-m,p,\delta} + \|Pu_\infty\|_{0,p,\delta}) \\ &\leq C_2(\|(1-\psi)Pu\|_{k-m,p,\delta} + \|[P,\psi]u\|_{k-m,p,\delta}) \\ &\leq C_3(\|Pu\|_{k-m,p,\delta} + \|u\|_{k-1,p;\overline{V}}),\end{aligned}$$

where in the last line we have used the fact that $[P,\psi]$ is an operator of order $m-1$ supported on $\overline{V}$.

For the second term, standard interior elliptic estimates yield

$$\begin{aligned}\|u_0\|_{k,p,\delta} &= \|u_0\|_{k,p,\delta;\overline{U}} \\ &\leq C_1(\|Pu_0\|_{k-m,p,\delta;\overline{V}} + \|u_0\|_{0,p,\delta;\overline{V}}) \\ &\leq C_1(\|\psi Pu\|_{k-m,p,\delta} + \|[P,\psi]u\|_{k-m,p,\delta} + \|u\|_{0,p,\delta;\overline{V}}) \\ &\leq C_2(\|Pu\|_{k-m,p,\delta} + \|u\|_{k-1,p;\overline{V}}).\end{aligned}$$

Finally, standard interpolation inequalities on the compact set $\overline{V}$ (see [**10**, Thm. 3.70]) allow us to replace $\|u\|_{k-1,p;\overline{V}}$ by $C\|u\|_{0,p;\overline{V}} + \varepsilon\|u\|_{k,p,\delta}$ with an arbitrarily small constant ε. Absorbing the ε term on the left-hand side, we conclude

$$\|u\|_{k,p,\delta} \leq C(\|Pu\|_{k-m,p,\delta} + \|u\|_{0,p;\overline{V}}). \tag{4.11}$$

Now suppose $\{u_i\}$ is a sequence in $\operatorname{Ker} P \cap H^{k,p}_\delta(M;E)$. Normalize u_i so that $\|u_i\|_{k,p,\delta} = 1$. By the Rellich lemma on the compact set $\overline{V}$, some subsequence converges in $H^{0,p}(\overline{V};E)$. From (4.11) we conclude that this subsequence is Cauchy and hence convergent in $H^{k,p}_\delta(M;E)$, and therefore $\operatorname{Ker} P \cap H^{k,p}_\delta(M;E)$ is finite-dimensional.

Next we need to show that P has closed range. Since $\operatorname{Ker} P$ is finite-dimensional, it has a closed complementary subspace $Y \subset H^{k,p}_\delta(M;E)$. I claim there is a constant C such that

$$\|u\|_{k,p,\delta} \leq C\|Pu\|_{k-m,p,\delta} \qquad \text{for all } u \in Y. \tag{4.12}$$

If not, there is a sequence $\{u_i\} \subset Y$ with

$$\|u_i\|_{k,p,\delta} = 1 \text{ and } \|Pu_i\|_{k-m,p,\delta} \to 0.$$

By the Rellich lemma again, there exists a subsequence (still denoted by $\{u_i\}$) that converges in $H^{0,p}(\overline{V};E)$. Then (4.11) shows that the subsequence also converges in $H^{k,p}_\delta(M;E)$. However, the limit u then satisfies $\|u\|_{k,p,\delta} = 1$ and $Pu = 0$ by continuity, and is therefore a nonzero element of $Y \cap \operatorname{Ker} P$, which is a contradiction.

Now if $u_i \in H^{k,p}_\delta(M;E)$ with $Pu_i = f_i \to f$ in $H^{k-m,p}_\delta(M;F)$, we can assume without loss of generality that each $u_i \in Y$, and then (4.12) shows that $\{u_i\}$ converges in $H^{k,p}_\delta(M;E)$, which shows that P has closed range.

Conversely, suppose $P\colon H^{k,p}_\delta(M;E) \to H^{k-m,p}_\delta(M;E)$ is semi-Fredholm. Let $K \subset M$ be a compact subset chosen as follows: If P has the unique continuation property, K can be any compact subset of M with nonempty interior; otherwise, let $K = M \smallsetminus A_\varepsilon$, where $\varepsilon > 0$ is chosen small enough that no element of the finite-dimensional space $\operatorname{Ker} P \cap H^{m,p}_\delta(M;E)$ vanishes identically on K.

The key fact is that there exists $c > 0$ such that

$$\|u - v\|_{0,p,\delta} \geq c\|u\|_{0,p,\delta} \tag{4.13}$$

whenever $u \in H^{0,p}_\delta(M;E)$ is supported in $M \smallsetminus K$ and $v \in \operatorname{Ker} P \cap H^{m,p}_\delta(M;E)$. To see this, note that our choice of K ensures that $\|\cdot\|_{0,p,\delta;K}$ is a norm on $\operatorname{Ker} P \cap H^{m,p}_\delta(M;E)$. Since all norms on a finite-dimensional vector space are equivalent, there exists a constant a such that

$$\|v\|_{0,p,\delta;K} \geq a\|v\|_{0,p,\delta}$$

for all $v \in \operatorname{Ker} P \cap H^{m,p}_\delta(M;E)$. Suppose now that u is supported in $M \smallsetminus K$, $\|u\|_{0,p,\delta} = 1$, and $v \in \operatorname{Ker} P \cap H^{m,p}_\delta(M;E)$. If $\|v\|_{0,p,\delta} \leq 1/(1+a)$, then by the reverse triangle inequality

$$\|u - v\|_{0,p,\delta} \geq \|u\|_{0,p,\delta} - \|v\|_{0,p,\delta} \geq 1 - \frac{1}{1+a} = \frac{a}{1+a}.$$

If on the other hand $\|v\|_{0,p,\delta} \geq 1/(1+a)$, then because u vanishes on K,

$$\|u - v\|_{0,p,\delta} \geq \|u - v\|_{0,p,\delta;K} = \|v\|_{0,p,\delta;K}$$
$$\geq a\|v\|_{0,p,\delta} \geq \frac{a}{1+a}.$$

Inequality (4.13) (with $c = a/(1+a)$) then follows for general u by homogeneity.

As above, $\operatorname{Ker} P \cap H^{k,p}_\delta(M;E)$ has a closed complementary subspace Y, and both Y and $P(H^{k,p}_\delta(M;E)) \subset H^{k-m,p}_\delta(M;E)$ are Banach spaces with the induced norms. Then $P|_Y \colon Y \to P(H^{k,p}_\delta(M;E))$ is bijective, and its inverse $(P|_Y)^{-1} \colon P(H^{k,p}_\delta(M;E)) \to Y$ is bounded by the open mapping theorem. This means there exists a constant $C > 0$ such that (4.12) holds.

Let $u \in H^{k,p}_\delta(M;E)$ be supported in $M \smallsetminus K$, and write $u = u_0 + u_Y$, with $u_0 \in \operatorname{Ker} P$, $u_Y \in Y$. It follows from (4.13) that

$$\|u_Y\|_{0,p,\delta} = \|u - u_0\|_{0,p,\delta} \geq c\|u\|_{0,p,\delta},$$

and therefore, by (4.12) with $k = m$,

$$\|u\|_{0,p,\delta} \leq c^{-1}\|u_Y\|_{0,p,\delta} \leq c^{-1}\|u_Y\|_{m,p,\delta}$$
$$\leq c^{-1}C\|Pu_Y\|_{0,p,\delta} = c^{-1}C\|Pu\|_{0,p,\delta},$$

which is (4.9).

Finally, suppose that both (4.9) and (4.10) hold. To show that P is actually Fredholm, all that remains to be shown is that the range of P has finite codimension in $H^{k-m,p}_\delta(M;F)$. Recall that $H^{0,p^*}_{-\delta}(M;E)$ is dual to $H^{0,p}_\delta(M;E)$ under the standard L^2 pairing. The argument above, using (4.10) instead of (4.9), shows that $P^* = P\colon H^{k,p^*}_{-\delta}(M;E) \to H^{k-m,p^*}_{-\delta}(M;E)$ has finite-dimensional kernel. Any $v \in H^{0,p^*}_{-\delta}(M;E)$ that annihilates the range of P in $H^{k-m,p}_\delta(M;E)$ satisfies in particular $(v, Pu) = 0$ for all $u \in C^\infty_c(M;E)$, so is a distribution solution to $Pv = 0$.

By Lemma 4.8, $v \in H^{k,p^*}_{-\delta}(M;E)$. Thus there is at most a finite-dimensional subspace of $H^{0,p^*}_{-\delta}(M;E)$ that annihilates the range of P. Since $P(H^{k,p}_\delta(M;E))$ is closed in $H^{k-m,p}_\delta(M;E)$, it has finite codimension. □

CHAPTER 5

Analysis on Hyperbolic Space

In this chapter, we will analyze the behavior of geometric elliptic operators on hyperbolic space, which serves as a model for the more general case. Our goal is to show that if P is an operator on hyperbolic space satisfying the hypotheses of Theorem C, then P is an isomorphism on appropriate weighted Sobolev and Hölder spaces. (In the next chapter, we will use the resulting inverse map to piece together a parametrix for the analogous operator acting on an arbitrary asymptotically hyperbolic manifold.)

For the purposes of this chapter, we will use the Poincaré ball model, identifying hyperbolic space with the unit ball $\mathbb{B} \subset \mathbb{R}^{n+1}$, with coordinates $(\xi^1, \ldots, \xi^{n+1})$, and with the hyperbolic metric $\breve{g} = 4(1 - |\xi|)^{-2} \sum_i (d\xi^i)^2$. The hyperbolic distance function can be written in terms of the Euclidean norm and dot product as

$$d_{\breve{g}}(\xi, \eta) = \cosh^{-1} \frac{(1 + |\xi|^2)(1 + |\eta|^2) - 4\xi \cdot \eta}{(1 - |\xi|^2)(1 - |\eta|^2)}.$$

It will be convenient to use

$$\rho(\xi) = \frac{1}{\cosh d_{\breve{g}}(\xi, 0)} = \frac{1 - |\xi|^2}{1 + |\xi|^2}$$

as a defining function for the ball, where $0 = (0, \ldots, 0)$ denotes the origin in $\mathbb{B} \subset \mathbb{R}^{n+1}$.

Throughout this chapter, E will be a geometric tensor bundle of weight r over $\mathbb{B}$, and $P \colon C^\infty(\mathbb{B}; E) \to C^\infty(\mathbb{B}; E)$ will be a formally self-adjoint geometric elliptic operator of order m. The fact that P is geometric implies that it is isometry invariant: If φ is any orientation-preserving hyperbolic isometry and u is any section of E, then

$$\varphi^*(Pu) = P(\varphi^* u). \tag{5.1}$$

We will assume that P satisfies (1.4). Then by Lemma 4.10, $P \colon H^{m,2}(\mathbb{B}; E) \to L^2(\mathbb{B}; E)$ is Fredholm. The next lemma shows that this is equivalent to being an isomorphism.

LEMMA 5.1. *Suppose $P \colon C^\infty(\mathbb{B}; E) \to C^\infty(\mathbb{B}; E)$ is a geometric elliptic operator of order m on $\mathbb{B}$. Then $P \colon H^{m,2}(\mathbb{B}; E) \to H^{0,2}(\mathbb{B}; F)$ is Fredholm if and only if it is an isomorphism.*

PROOF. If P is an isomorphism, then clearly it is Fredholm. Conversely, if P is Fredholm, then by Lemma 4.10 P satisfies

$$\|u\| \leq C\|Pu\| \tag{5.2}$$

whenever u is supported in the complement of some compact set K. Suppose u is *any* smooth, compactly supported section of E. There is a Möbius transformation φ such that $\varphi^{-1}(\operatorname{supp} u) \subset \mathbb{B} \smallsetminus K$, so $\varphi^* u$ satisfies (5.2). Because P and the

L^2 norm are preserved by φ, u itself satisfies the same estimate. Therefore, by continuity, (5.2) holds for all $u \in H^{m,2}$, so $\operatorname{Ker} P$ is trivial. Since P is self-adjoint as an unbounded operator on $L^2(M; E)$, its index is zero, which means that it is also surjective. □

Let K be the Green kernel of P: That is, K is the Schwartz kernel of the operator $P^{-1}\colon L^2(\mathbb{B}; E) \to H^{m,2}(\mathbb{B}; E)$. Invariantly, K is interpreted as a distributional section of the bundle $\operatorname{Hom}(\pi_2^* E, \pi_1^* E)$ over $\mathbb{B} \times \mathbb{B}$, where π_j is projection on the jth factor. For all $f \in L^2(\mathbb{B}; E)$,

$$P^{-1}f(\xi) = \int_{\mathbb{B}} K(\xi, \eta) f(\eta) dV_{\breve{g}}(\eta). \tag{5.3}$$

Equivalently, if we write $K_\eta(\xi) = K(\xi, \eta)$, K_η satisfies

$$PK_\eta(\xi) = \delta_\eta(\xi) \operatorname{Id}_{E_\eta}$$

in the distribution sense and $K_\eta \in L^2$ on the complement of a neighborhood of η. Somewhat more explicitly, for any $\eta \in \mathbb{B}$ and $u_0 \in E_\eta$, $K_\eta u_0$ is a (distributional) section of E satisfying $P(K_\eta u_0) = \delta_\eta u_0$. In particular, $K_0(\xi) = K(\xi, 0)$ can be viewed as a fundamental solution for P on $\mathbb{B}$ with pole at 0.

By local elliptic regularity, K is C^∞ away from the diagonal $\{\xi = \eta\}$. Since P is formally self-adjoint, it is easy to check that K satisfies the symmetry condition $K(\eta, \xi) = K(\xi, \eta)^*$ for $\xi \neq \eta$, where $K(\xi, \eta)^*\colon E_\xi \to E_\eta$ is the (pointwise) adjoint of $K(\xi, \eta) \in \operatorname{Hom}(E_\eta, E_\xi)$.

We extend our defining function ρ to a function $\rho\colon \overline{\mathbb{B}} \times \overline{\mathbb{B}} \to [0, 1]$ of two variables (still denoted by the same symbol) by

$$\rho(\xi, \eta) = \frac{1}{\cosh d_{\breve{g}}(\xi, \eta)} = \frac{(1 - |\xi|^2)(1 - |\eta|^2)}{(1 + |\xi|^2)(1 + |\eta|^2) - 4\xi \cdot \eta}.$$

Observe that $\rho(\xi, 0) = \rho(\xi)$.

Our main technical tool in this chapter is the following decay estimate for K.

PROPOSITION 5.2. *Let $P\colon C^\infty(\mathbb{B}; E) \to C^\infty(\mathbb{B}; E)$ be a formally self-adjoint geometric elliptic operator of order m satisfying* (1.4). *Then P has positive indicial radius R, and for any $\varepsilon > 0$ there is a constant C such that*

$$|K(\xi, \eta)| \leq C\rho(\xi, \eta)^{n/2 + R - \varepsilon} \tag{5.4}$$

whenever $d_{\breve{g}}(\xi, \eta) \geq 1$. (*The norm here is the pointwise operator norm on $\operatorname{Hom}(E_\eta, E_\xi)$ with respect to the hyperbolic metric.*)

PROOF. The isometry invariance of P implies that K has the following equivariance property for any orientation-preserving hyperbolic isometry φ:

$$K(\varphi(\xi), \varphi(\eta)) = (\varphi^*)^{-1} \circ K(\xi, \eta) \circ \varphi^*. \tag{5.5}$$

Also, since ρ is defined purely in terms of the hyperbolic distance function, it is clearly isometry invariant:

$$\rho(\varphi(\xi), \varphi(\eta)) = \rho(\xi, \eta).$$

Therefore, to prove (5.4), it suffices to show that

$$|K(\xi, 0)| \leq C\rho(\xi, 0)^{n/2 + R - \varepsilon} = C\rho(\xi)^{n/2 + R - \varepsilon}.$$

Note that $K_0(\xi) = K(\xi, 0)$ defines a smooth section of the bundle $\operatorname{Hom}(E_0, E)$ over $\mathbb{B} \smallsetminus \{0\}$, whose fiber at ξ is the vector space $\operatorname{Hom}(E_0, E_\xi)$.

The group of orientation-preserving hyperbolic isometries that fix 0 is exactly $\mathrm{SO}(n+1)$, acting linearly on the unit ball as isometries of both the hyperbolic and Euclidean metrics. Let $L \subset \mathbb{B}$ be the ray $L = \{(0,\ldots,0,t) : 0 \le t < 1\}$ along the ξ^{n+1}-axis. The subgroup of $\mathrm{SO}(n+1)$ that fixes L pointwise is $\mathrm{SO}(n) \subset \mathrm{SO}(n+1)$, realized as the group of linear isometries acting in the first n variables only. Observe that for each $\xi_0 \in L$, $\mathrm{SO}(n)$ acts orthogonally (or unitarily if E is a complex tensor bundle) on the fiber E_{ξ_0} by pulling back.

Let $E_0 = E_0^{(1)} \oplus \cdots \oplus E_0^{(k)}$ be an orthogonal decomposition of E_0 into irreducible $\mathrm{SO}(n)$-invariant subspaces. We extend this to a decomposition of the bundle E over $\mathbb{B} \smallsetminus \{0\}$ as follows. First, for each $\xi_0 \in L$, let $E_{\xi_0}^{(i)}$ be the subspace of E_{ξ_0} obtained by parallel translating $E_0^{(i)}$ along L with respect to the Euclidean metric $\bar{g}$; since $\mathrm{SO}(n)$ acts as Euclidean isometries, it follows that

$$E_{\xi_0} = E_{\xi_0}^{(1)} \oplus \cdots \oplus E_{\xi_0}^{(k)}$$

is an orthogonal irreducible $\mathrm{SO}(n)$-decomposition of E_{ξ_0}. Then for an arbitrary point $\xi \in \mathbb{B} \smallsetminus \{0\}$, let $E_\xi^{(i)} = (\varphi^*)^{-1} E_{\xi_0}^{(i)}$, where ξ_0 is the unique point of L such that $|\xi_0| = |\xi|$, and $\varphi \in \mathrm{SO}(n+1)$ satisfies $\varphi(\xi_0) = \xi$. Since $E_{\xi_0}^{(i)}$ is invariant under $\mathrm{SO}(n)$, $E_\xi^{(i)}$ does not depend on the choice of φ. Since φ can be chosen locally to depend smoothly on ξ (by means of a smooth local section of the submersion $\mathrm{SO}(n+1) \to \mathrm{SO}(n+1)/\mathrm{SO}(n) = \mathbb{S}^n$), this results in k smooth subbundles $E^{(1)},\ldots,E^{(k)}$ of E over $\mathbb{B} \smallsetminus \{0\}$.

For each pair of indices $i,j = 1,\ldots,k$, we choose an $\mathrm{SO}(n)$-equivariant linear map $k_0^{(i,j)} \colon E_0 \to E_0$ as follows: If $E_0^{(i)}$ and $E_0^{(j)}$ are isomorphic as representations of $\mathrm{SO}(n)$, let $k_0^{(i,j)}$ be an $\mathrm{SO}(n)$-equivariant Euclidean isometry from $E_0^{(i)}$ to $E_0^{(j)}$, extended to be zero on $E_0^{(l)}$ for $l \ne i$; and otherwise let $k_0^{(i,j)}$ be the zero map. By Schur's lemma, the nonzero maps $k_0^{(i,j)}$ form a basis for the space of $\mathrm{SO}(n)$-equivariant endomorphisms of E_0. Let us renumber these nonzero maps as $k_0^1, \ldots, k_0^N$.

Next we extend each map k_0^j to a section k^j of $\mathrm{Hom}(E_0, E)$ over $\mathbb{B} \smallsetminus \{0\}$ in the same way as we extended the spaces $E_0^{(i)}$: First, for each point $\xi_0 \in L$, define $k_{\xi_0}^j \colon E_0 \to E_{\xi_0}$ to be k_0^j followed by $\bar{g}$-parallel translation along L from 0 to ξ_0; and then for arbitrary $\xi \in \mathbb{B} \smallsetminus \{0\}$, define $k_\xi^j = (\varphi^*)^{-1} \circ k_{\xi_0}^j \circ \varphi^*$, where $\xi_0 \in L$ and $\varphi \in \mathrm{SO}(n+1)$ satisfies $\varphi(\xi_0) = \xi$.

I claim that there are smooth functions $f_1, \ldots, f_N \colon (0,1) \to \mathbb{C}$ such that

$$K_0(\xi) = \sum_j f_j(\rho(\xi)) k_\xi^j \tag{5.6}$$

for all $\xi \in \mathbb{B} \smallsetminus \{0\}$. To see this, first note that for any point $\xi_0 \in L$, the equivariance property (5.5) implies that $K_0(\xi_0)$ is an $\mathrm{SO}(n)$-equivariant linear map from E_0 to E_{ξ_0}, and therefore by Schur's lemma it can be written as a linear combination of the maps $k_{\xi_0}^j$:

$$K_0(\xi_0) = \sum_j c_j(\xi_0) k_{\xi_0}^j.$$

For any other point $\xi \in \mathbb{B} \smallsetminus \{0\}$, let ξ_0 be the point of L such that $|\xi_0| = |\xi|$, and let $\varphi \in \mathrm{SO}(n+1)$ satisfy $\varphi(\xi_0) = \xi$. Then (5.5) yields

$$\begin{aligned} K_0(\xi) &= K(\varphi(\xi_0), \varphi(0)) \\ &= (\varphi^*)^{-1} \circ K(\xi_0, 0) \circ \varphi^* \\ &= \sum_j c_j(\xi_0)(\varphi^*)^{-1} \circ k_{\xi_0}^j \circ \varphi^* \\ &= \sum_j c_j(\xi_0) k_\xi^j. \end{aligned} \qquad (5.7)$$

Since $\rho\colon L \smallsetminus \{0\} \to (0,1)$ is a diffeomorphism, there are functions $f_j\colon (0,1) \to \mathbb{C}$ such that $f_j(\rho(\xi)) = f_j(\rho(\xi_0)) = c_j(\xi_0)$ whenever $|\xi| = |\xi_0|$, and then (5.7) is equivalent to (5.6). The smoothness of K_0 implies that the functions f_j are smooth.

Now the equation $PK_0 = 0$ reduces to an analytic system of ordinary differential equations for the functions f_j, and the fact that P is uniformly degenerate implies that this system has a regular singular point at $\rho = 0$. Because the sections k^j of $\mathrm{Hom}(E_0, E)$ extend smoothly to $\overline{\mathbb{B}} \smallsetminus \{0\}$, our definition of the indicial map of P guarantees that the characteristic exponents of this system of ODEs are precisely the characteristic exponents of the operator P. Therefore, by the standard theory of ODEs with regular singular points, each coefficient function f_j satisfies

$$|f_j(t)| \sim C_j t^{s_j} |\log t|^{k_j} \text{ as } t \to 0 \qquad (5.8)$$

for some characteristic exponent s_j and some nonnegative integer k_j.

If u_0 is any tensor in one of the summands $E_0^{(i)}$, then the images $k_\xi^j(u_0)$ lie in different summands of E_ξ and are therefore orthogonal, so

$$|K_0(\xi) u_0|_{\bar g}^2 = \sum_j |f_j(\rho(\xi))|^2 |k_\xi^j u_0|_{\bar g}^2,$$

where $\bar g$ is the Euclidean metric on $\mathbb{B}$. Since k_ξ^j is a Euclidean isometry onto its image, $|k_\xi^j u_0|_{\bar g}$ is independent of ξ. Therefore, if $k_\xi^j u_0 \ne 0$, then on $\mathbb{B} \smallsetminus \{0\}$ we have

$$|K_0(\xi) u_0|_{\bar g} \ge C|f_j(\rho(\xi))| \ge C\rho(\xi)^{\mathrm{Re}\, s_j} |\log \rho(\xi)|^{k_j}$$

for some positive constant C. Because $K_0(\xi) u_0$ is in L^2 away from 0, by Lemma 3.2 we must have $\mathrm{Re}\, s_j > n/2 - r$ for each such j. Since for each j there is some u_0 such that $k_\xi^j u_0 \ne 0$, the same inequality holds for every j. By definition of the indicial radius R, this implies that in fact $R > 0$ and $\mathrm{Re}\, s_j \ge n/2 - r + R$. Using (5.8), we conclude that for any $\varepsilon > 0$ there is a constant C such that

$$|f_j(t)| \le C t^{n/2-r+R-\varepsilon} \text{ for } t \text{ away from 1.}$$

This in turn implies

$$\begin{aligned} |K_0(\xi) u_0|_{\breve g} &\le C \rho(\xi)^r |K_0(\xi) u_0|_{\bar g} \\ &\le C' \rho(\xi)^r \rho(\xi)^{n/2-r+R-\varepsilon} |u_0|_{\bar g} \\ &= C'' \rho(\xi)^{n/2+R-\varepsilon} |u_0|_{\breve g} \end{aligned}$$

whenever $d(\xi, 0) \ge 1$, which was to be proved. $\square$

The next two lemmas give some estimates that will be needed to use our decay estimate for proving mapping properties of P^{-1}.

LEMMA 5.3. *For any real numbers p, q, r such that $p+1 > 0$ and $r > q+1 > 0$, there exists a constant C depending only on p, q, r such that the following estimate holds for all $u \in [0, 1)$:*

$$\int_0^1 \frac{t^p(1-t)^q}{(1-ut)^r} \, dt \leq C(1-u)^{q+1-r}.$$

PROOF. We use the following standard integral representation for hypergeometric functions [**27**, p. 59]:

$$F(\alpha, \beta, \gamma; z) = \frac{\Gamma(\gamma)}{\Gamma(\beta)\Gamma(\gamma-\beta)} \int_0^1 \frac{t^{\beta-1}(1-t)^{\gamma-\beta-1}}{(1-tz)^\alpha} \, dt,$$

which is valid if $\operatorname{Re}\gamma > \operatorname{Re}\beta > 0$ and $|z| < 1$. The hypergeometric function $F(\alpha, \beta, \gamma; z)$ is analytic for $|z| < 1$ and satisfies a second-order ODE that has a regular singular point at $z = 1$ with characteristic exponents 0 and $\gamma - \alpha - \beta$ [**17**, p. 246]. As long as $\gamma - \alpha - \beta < 0$, therefore, it satisfies

$$|F(\alpha, \beta, \gamma; u)| \leq C(1-u)^{\gamma-\alpha-\beta} \quad \text{if } 0 \leq u < 1.$$

Applying this with $\alpha = r$, $\beta = p+1$, and $\gamma = p+q+2$ proves the lemma. $\square$

LEMMA 5.4. *Suppose a and b are real numbers such that $a + b > n$ and $a > b$. There exists a constant C depending only on n, a, b such that the following estimate holds for all $\xi, \zeta \in \mathbb{B}$:*

$$\int_{\mathbb{B}} \rho(\xi, \eta)^a \rho(\eta, \zeta)^b \, dV_{\breve{g}}(\eta) \leq C\rho(\xi, \zeta)^b.$$

PROOF. By an isometry, we can arrange that $\zeta = 0$ and $\xi = (0, \ldots, 0, r)$ is on the positive ξ^{n+1}-axis. Substituting $\zeta = 0$ into the integral, we must estimate

$$I := \int_{\mathbb{B}} \left(\frac{(1-|\xi|^2)(1-|\eta|^2)}{(1+|\xi|^2)(1+|\eta|^2) - 4\xi \cdot \eta} \right)^a \left(\frac{1-|\eta|^2}{1+|\eta|^2} \right)^b dV_{\breve{g}}(\eta). \qquad (5.9)$$

Parametrize the ball by the map $\Phi \colon (0, 1) \times (0, \pi) \times \mathbb{S}^{n-1} \to \mathbb{B}$ given by

$$\Phi(s, \theta, \omega) = (s\omega^1 \sin\theta, \ldots, s\omega^n \sin\theta, s\cos\theta),$$

so that s is the Euclidean distance from 0 and θ is the angle from the positive ξ^{n+1}-axis. In these coordinates, the hyperbolic metric is

$$\breve{g} = \frac{4}{(1-s^2)^2}(ds^2 + s^2 d\theta^2 + s^2 \sin^2\theta \, \mathring{g}),$$

where $\mathring{g}$ represents the standard metric on $\mathbb{S}^{n-1}$. The hyperbolic volume element is therefore

$$dV_{\breve{g}} = \frac{2^{n+1} s^n \sin^{n-1}\theta}{(1-s^2)^{n+1}} \, ds \, d\theta \, dV_{\mathring{g}},$$

where $dV_{\mathring{g}}$ is the volume element on $\mathbb{S}^{n-1}$.

In these coordinates, we have $|\xi| = r$, $|\eta| = s$ and $\xi \cdot \eta = rs \cos\theta$. The integrand in (5.9) is constant on each sphere $\mathbb{S}^{n-1}$, so we can immediately integrate over $\mathbb{S}^{n-1}$ and write I as a constant multiple of

$$\int_0^1 \int_0^\pi \left(\frac{(1-r^2)(1-s^2)}{(1+r^2)(1+s^2) - 4rs\cos\theta} \right)^a \left(\frac{1-s^2}{1+s^2} \right)^b \frac{s^n \sin^{n-1}\theta}{(1-s^2)^{n+1}} \, d\theta \, ds.$$

Since we are only interested in estimates up to a constant multiple, we will write $f \sim g$ to mean that f/g is bounded above and below by positive constants depending only on a, b, and n. Thus, for example, $1 - s^2 = (1-s)(1+s) \sim 1 - s$, $1 - r^2 \sim 1 - r$, $1 + r^2 \sim 1 + s^2 \sim 1$, and

$$I \sim \int_0^1 \int_0^\pi \left(\frac{(1-r)(1-s)}{(1+r^2)(1+s^2) - 4rs\cos\theta}\right)^a (1-s)^b \frac{s^n \sin^{n-1}\theta}{(1-s)^{n+1}} d\theta\, ds. \tag{5.10}$$

The θ integral can be simplified by the substitution $\cos\theta = 2t - 1$ to obtain

$$\int_0^\pi \frac{\sin^{n-1}\theta}{((1+r^2)(1+s^2) - 4rs\cos\theta)^a} d\theta \tag{5.11}$$
$$= 2^{n-1} B(r,s)^a \int_0^1 \frac{t^{n/2-1}(1-t)^{n/2-1}}{(1 - 8B(r,s)rst)^a} dt,$$

where

$$B(r,s) = \frac{1}{(1+r^2)(1+s^2) + 4rs} \sim 1.$$

Because our hypothesis guarantees that $n - a < b < a$ and therefore $a > n/2$, Lemma 5.3 shows that the right-hand integral in (5.11) is bounded by a constant multiple of $(1 - 8B(r,s)rs)^{n/2-a}$. Substituting this into (5.10) yields

$$I \leq C(1-r)^a \int_0^1 \frac{s^n(1-s)^{a+b-n-1}}{(1 - 8B(r,s)rs)^{a-n/2}} ds. \tag{5.12}$$

A computation shows that

$$1 - 8B(r,s)rs = \frac{(1-rs)^2 + (r-s)^2}{(1+rs)^2 + (r+s)^2} \sim (1-rs)^2 + (r-s)^2 \geq (1-rs)^2.$$

Inserting this into (5.12), we conclude that

$$I \leq C(1-r)^a \int_0^1 \frac{s^n(1-s)^{a+b-n-1}}{(1-rs)^{2a-n}} ds.$$

Lemma 5.3 then shows that this is bounded by a multiple of $(1-r)^b \sim \rho(\xi,\zeta)^b$. $\square$

The following estimate is the key to proving sharp mapping properties of P^{-1}.

LEMMA 5.5. *Let P satisfy the hypotheses of Proposition 5.2. Then for any real number b satisfying $n/2 - R < b < n/2 + R$, there exists a constant C such that*

$$\int_\mathbb{B} |K(\xi,\eta)|\, \rho(\eta)^b\, dV_{\breve{g}}(\eta) \leq C\rho(\xi)^b,$$
$$\int_\mathbb{B} |K(\xi,\eta)|\, \rho(\xi)^b\, dV_{\breve{g}}(\xi) \leq C\rho(\eta)^b.$$

PROOF. Since $|K(\xi,\eta)| = |K(\eta,\xi)^*| = |K(\eta,\xi)|$ by self-adjointness, the two inequalities are equivalent, so it suffices to prove the second one. We will write

$$\int_\mathbb{B} |K(\xi,\eta)|\, \rho(\xi)^b\, dV_{\breve{g}}(\xi)$$
$$= \int_{d_{\breve{g}}(\xi,\eta) \leq 1} |K(\xi,\eta)|\, \rho(\xi)^b\, dV_{\breve{g}}(\xi) + \int_{d_{\breve{g}}(\xi,\eta) \geq 1} |K(\xi,\eta)|\, \rho(\xi)^b\, dV_{\breve{g}}(\xi)$$

and estimate each term separately.

For the first term, we observe that K is uniformly locally integrable near $\xi = \eta$: Since $K_0(\xi) = K(\xi, 0)$ satisfies $PK_0 = \delta_0 \operatorname{Id}_{E_0}$, and the Dirac delta function is in the Sobolev space $H^{-1,q}$ for $1 < q < 1 + 1/n$ (defined as the dual space to H^{1,q^*}, $q^* = q/(q-1)$), local elliptic regularity implies that $K_0 \in L^q_{\text{loc}} \subset L^1_{\text{loc}}$. Therefore, if $\eta \in \mathbb{B}$ is arbitrary and φ is any Möbius transformation sending η to 0, the change of variables $\xi' = \varphi(\xi)$ yields

$$\int_{d_{\breve{g}}(\xi,\eta) \leq 1} |K(\xi,\eta)|\, dV_{\breve{g}}(\xi) = \int_{d_{\breve{g}}(\xi',0) \leq 1} |K(\xi',0)|\, dV_{\breve{g}}(\xi') \leq C. \tag{5.13}$$

Using the triangle inequality together with the elementary fact that $\cosh(A+B) \leq 2 \cosh A \cosh B$ for $A, B \geq 0$, we estimate

$$\cosh d_{\breve{g}}(\xi, 0) \leq \cosh(d_{\breve{g}}(\xi, \eta) + d_{\breve{g}}(\eta, 0)) \leq 2 \cosh d_{\breve{g}}(\xi, \eta) \cosh d_{\breve{g}}(\eta, 0).$$

It follows that $\rho(\eta) \leq 2\rho(\xi)$ on the set where $d_{\breve{g}}(\xi, \eta) \leq 1$. By symmetry, the same inequality holds with ξ and η reversed. Thus

$$\int_{d_{\breve{g}}(\xi,\eta) \leq 1} |K(\xi,\eta)| \rho(\xi)^b\, dV_{\breve{g}}(\xi)$$
$$\leq \left(\sup_{d_{\breve{g}}(\xi,\eta) \leq 1} \rho(\xi)^b \right) \int_{d_{\breve{g}}(\xi,\eta) \leq 1} |K(\xi,\eta)|\, dV_{\breve{g}}(\xi)$$
$$\leq C\rho(\eta)^b.$$

For the second term, choose $\varepsilon > 0$ small enough that

$$\frac{n}{2} - R + \varepsilon < b < \frac{n}{2} + R - \varepsilon. \tag{5.14}$$

Then with $a = n/2 + R - \varepsilon$, we have $a + b > n$ and $a > b$, so we can use Proposition 5.2 and Lemma 5.4 to conclude

$$\int_{d_{\breve{g}}(\xi,\eta) \geq 1} |K(\xi,\eta)| \rho(\xi)^b\, dV_{\breve{g}}(\xi) \leq \int_{d_{\breve{g}}(\xi,\eta) \geq 1} \rho(\xi,\eta)^a \rho(\xi)^b\, dV_{\breve{g}}(\xi)$$
$$\leq C\rho(\xi)^b.$$

$\square$

PROPOSITION 5.6. *If $1 < p < \infty$, $k \geq m$, and $|\delta + n/p - n/2| < R$, then there exists a constant C such that*

$$\|u\|_{k,p,\delta} \leq C\|Pu\|_{k-m,p,\delta} \tag{5.15}$$

for all $u \in H^{k,p}_\delta(\mathbb{B}; E)$.

PROOF. Using Lemma 4.8, it suffices to prove that

$$\|u\|_{0,p,\delta} \leq C\|Pu\|_{0,p,\delta}$$

for all $u \in H^{k,p}_\delta(\mathbb{B}; E)$. Because $C^\infty_c(\mathbb{B}; E)$ is dense in $H^{k,p}_\delta(\mathbb{B}; E)$, it suffices to prove this inequality for $u \in C^\infty_c(\mathbb{B}; E)$. Since $u = P^{-1}(Pu)$ in that case, it suffices to prove the estimate

$$\|P^{-1}f\|_{0,p,\delta} \leq C\|f\|_{0,p,\delta} \text{ for all } f \in C^\infty_c(\mathbb{B}; E). \tag{5.16}$$

Put
$$p^* = \frac{p}{p-1},$$
$$a = \frac{1}{p^*}\left(\delta + \frac{n}{p}\right),$$

so that
$$\begin{aligned}\frac{n}{2} - R < ap^* < \frac{n}{2} + R,\\ \frac{n}{2} - R < ap - \delta p < \frac{n}{2} + R.\end{aligned} \qquad (5.17)$$

By Hölder's inequality and Lemma 5.5, we estimate

$$\begin{aligned}|P^{-1}f(\xi)|_{\check g} &\leq \int_{\mathbb{B}} |K(\xi,\eta)|\,|f(\eta)|_{\check g}\,dV_{\check g}(\eta)\\ &= \int_{\mathbb{B}} \left(|K(\xi,\eta)|^{1/p}\rho(\eta)^{-a}|f(\eta)|_{\check g}\right)\left(|K(\xi,\eta)|^{1/p^*}\rho(\eta)^{a}\right)dV_{\check g}(\eta)\\ &\leq \left(\int_{\mathbb{B}}|K(\xi,\eta)|\,\rho(\eta)^{-ap}|f(\eta)|_{\check g}^{p}\,dV_{\check g}(\eta)\right)^{1/p}\times\\ &\qquad\left(\int_{\mathbb{B}}|K(\xi,\eta)|\,\rho(\eta)^{ap^*}\,dV_{\check g}(\eta)\right)^{1/p^*}\\ &\leq C\rho(\xi)^a\left(\int_{\mathbb{B}}|K(\xi,\eta)|\,\rho(\eta)^{-ap}|f(\eta)|_{\check g}^{p}\,dV_{\check g}(\eta)\right)^{1/p}.\end{aligned}$$

Therefore,
$$\begin{aligned}\|P^{-1}f\|_{0,p,\delta}^p &= \int_{\mathbb{B}}\rho(\xi)^{-\delta p}|P^{-1}f(\xi)|_{\check g}^{p}\,dV_{\check g}(\xi)\\ &\leq C^p \int_{\mathbb{B}}\int_{\mathbb{B}}\rho(\xi)^{ap-\delta p}|K(\xi,\eta)|\,\rho(\eta)^{-ap}|f(\eta)|_{\check g}^{p}\,dV_{\check g}(\eta)\,dV_{\check g}(\xi).\end{aligned}$$

By Lemma 5.5 again, we can evaluate the ξ integral first to obtain
$$\begin{aligned}\|P^{-1}f\|_{0,p,\delta}^p &\leq C'\int_{\mathbb{B}}\rho(\eta)^{ap-\delta p}\rho(\eta)^{-ap}|f(\eta)|_{\check g}^{p}\,dV_{\check g}(\eta)\\ &= C'\|f\|_{0,p,\delta}^p.\end{aligned}$$

$\square$

THEOREM 5.7. *Let $P\colon C^\infty(\mathbb{B};E) \to C^\infty(\mathbb{B};E)$ be a formally self-adjoint geometric elliptic operator of order m satisfying (1.4). If $k \geq m$, $1 < p < \infty$, and $|\delta + n/p - n/2| < R$, then the natural extension $P\colon H^{k,p}_\delta(\mathbb{B};E) \to H^{k-m,p}_\delta(\mathbb{B};E)$ is an isomorphism.*

PROOF. Injectivity is an immediate consequence of (5.15). To prove surjectivity, let $f \in H^{k-m,p}_\delta(\mathbb{B};E)$ be arbitrary, and let $f_i \in C^\infty_c(\mathbb{B};E)$ be a sequence such that $f_i \to f$ in $H^{k-m,p}_\delta(\mathbb{B};E)$. Set $u_i = P^{-1}f_i \in H^{m,2}(\mathbb{B};E)$, so that $Pu_i = f_i$. Then each u_i is in $H^{0,p}_\delta(\mathbb{B};E)$ by (5.16), and in $H^{k,p}_\delta(\mathbb{B};E)$ by Lemma 4.8, and (5.15) shows that $\{u_i\}$ is Cauchy in $H^{k,p}_\delta(\mathbb{B};E)$. It follows that $u = \lim u_i \in H^{k,p}_\delta(\mathbb{B};E)$ satisfies $Pu = f$ as desired, so P is surjective. The continuity of the inverse map then follows from (5.15). $\square$

Now we turn our attention to the Hölder case. First we prove an estimate analogous to (5.16).

PROPOSITION 5.8. *If $|\delta - n/2| < R$, there exists a constant C such that*
$$\|P^{-1}f\|_{0,0,\delta} \leq C\|f\|_{0,0,\delta} \tag{5.18}$$
for all $f \in C_\delta^{0,0}(\mathbb{B}; E)$.

PROOF. By Lemma 5.5,
$$|P^{-1}f(\xi)|_{\breve{g}} \leq \int_{\mathbb{B}} |K(\xi,\eta)|\, |f(\eta)|_{\breve{g}}\, dV_{\breve{g}}(\eta)$$
$$\leq C \int_{\mathbb{B}} |K(\xi,\eta)|\, \rho(\eta)^\delta\, \|f\|_{0,0,\delta}\, dV_{\breve{g}}(\eta)$$
$$\leq C'\rho(\xi)^\delta \|f\|_{0,0,\delta},$$
which implies
$$\|P^{-1}f\|_{0,0,\delta} = \sup_{\xi \in \mathbb{B}} \left(\rho(\xi)^{-\delta} |P^{-1}f(\xi)|_{\breve{g}}\right) \leq C' \|f\|_{0,0,\delta}.$$
□

THEOREM 5.9. *Let $P\colon C^\infty(\mathbb{B}; E) \to C^\infty(\mathbb{B}; E)$ be a formally self-adjoint geometric elliptic operator of order m satisfying (1.4). If $0 < \alpha < 1$, $k \geq m$, and $|\delta - n/2| < R$, then the natural extension $P\colon C_\delta^{k,\alpha}(\mathbb{B}; E) \to C_\delta^{k-m,\alpha}(\mathbb{B}; E)$ is an isomorphism.*

PROOF. To prove surjectivity, let $f \in C_\delta^{k-m,\alpha}(\mathbb{B}; E)$ be arbitrary and set $u = P^{-1}f$, so that $u \in C_\delta^{0,0}(\mathbb{B}; E)$ by Proposition 5.8. An easy computation shows that $Pu = f$ in the distribution sense, so $u \in C_\delta^{k,\alpha}(\mathbb{B}; E)$ by Lemma 4.8.

To prove injectivity, choose δ' close to δ and p large such that $\delta > \delta' + n/p$ and $|\delta' + n/p - n/2| < R$. Then $C_\delta^{k,\alpha}(\mathbb{B}; E) \subset H_{\delta'}^{k,p}(\mathbb{B}; E)$ by by Lemma 3.6. Since P is injective on $H_{\delta'}^{k,p}(\mathbb{B}; E)$ by Theorem 5.7, it is injective on the smaller space $C_\delta^{k,\alpha}(\mathbb{B}; E)$. □

CHAPTER 6

Fredholm Theorems

In this chapter, we return to the general case of a connected $(n+1)$-manifold (M,g), assumed to be asymptotically hyperbolic of class $C^{l,\beta}$, with $l \geq 2$ and $0 \leq \beta < 1$. Let E be a geometric tensor bundle over M, and let $P\colon C^\infty(M;E) \to C^\infty(M;E)$ be a formally self-adjoint geometric elliptic operator of order $m \geq 1$. We will prove Fredholm properties of P by using Möbius coordinates near the boundary to piece together a parametrix modeled on the inverse operator on hyperbolic space.

For this purpose, we will need a slightly modified version of Möbius coordinates. Whereas the original Möbius coordinates defined in Chapter 2 were valid in a neighborhood of an interior point, for our parametrix construction we will need coordinates that are defined all the way up to the boundary and adjusted to make the background metric $\overline{g}$ close to the Euclidean metric on a neighborhood of a boundary point. These coordinates will be used to transfer the operator $\breve{P}^{-1}$ to M with an error that decays to one higher order along the boundary.

For each point $\widehat{p} \in \partial M$, choose some neighborhood Ω on which background coordinates (θ, ρ) are defined on a set of the form (2.1). Let $\omega^1, \ldots, \omega^n \in C^{l,\beta}_{(0)}(\Omega, T^*\overline{M})$ be 1-forms chosen so that $(\omega^1, \ldots, \omega^n, d\rho)$ is an orthonormal coframe for $\overline{g}$ at each point of $\partial M \cap \Omega$ (recall that $|d\rho|_{\overline{g}} \equiv 1$ along ∂M). Let A^α_β, B^α be the coefficients of ω^α at $\widehat{p}$, defined by

$$\omega^\alpha_{\widehat{p}} = A^\alpha_\beta d\theta^\beta_{\widehat{p}} + B^\alpha d\rho_{\widehat{p}},$$

and let $(\widetilde{\theta}^1, \ldots, \widetilde{\theta}^n)$ be the functions defined on Ω by

$$\widetilde{\theta}^\alpha = A^\alpha_\beta \theta^\beta + B^\alpha \rho.$$

Then $(\widetilde{\theta}^1, \ldots, \widetilde{\theta}^n, \rho)$ form coordinates on Ω, and in these new coordinates $\overline{g}$ has the matrix δ_{ij} at $\widehat{p}$. For $0 < a$ and $0 < r < c$, define open subsets $Y_a \subset \mathbb{H}$ and $Z_r(\widehat{p}) \subset \Omega \subset M$ by

$$Y_a = \{(x,y) \in \mathbb{H} : |x| < a, 0 < y < a\},$$
$$Z_r(\widehat{p}) = \{(\widetilde{\theta}, \rho) \in \Omega : |\widetilde{\theta}| < r, 0 < \rho < r\}.$$

For $0 < r < c$, define a chart $\Psi_{\widehat{p},r}\colon Y_1 \to Z_r(\widehat{p})$ by

$$(\widetilde{\theta}, \rho) = \Psi_{\widehat{p},r}(x,y) = (rx, ry).$$

We will call $\Psi_{\widehat{p},r}$ a *boundary Möbius chart* of radius r centered at $\widehat{p}$. Recall that $\breve{g}$ denotes the hyperboolic metric on the upper half-space.

LEMMA 6.1. *There is a constant $C > 0$ such that for any $\widehat{p} \in \partial M$ and any sufficiently small $r > 0$,*

$$\|\Psi^*_{\widehat{p},r} g - \breve{g}\|_{l,\beta;Y_1} \leq rC. \tag{6.1}$$

PROOF. Because Y_1 is not precompact in $\mathbb{H}$, we have to interpret the $C^{l,\beta}$ norm on the left-hand side of (6.1) as an intrinsic Hölder norm, defined by (3.2). For each point $(x_0, y_0) \in Y_1$, we have a Möbius chart $\Phi_{(x_0,y_0)} \colon B_2 \to V_2(x_0, y_0) \subset \mathbb{H}$ defined by

$$\Phi_{(x_0,y_0)}(x, y) = (x_0 + y_0 x, y_0 y).$$

Then we need to get an upper bound for

$$\sup_{(x_0,y_0) \in Y_1} \|\Phi^*_{(x_0,y_0)}(\Psi^*_{\widehat{p},r} g - \breve{g})\|_{C^{l,\beta}(B_2)}.$$

Since $\Phi_{(x_0,y_0)}$ is a hyperbolic isometry, the norm above is the same as

$$\|(\Psi_{\widehat{p},r} \circ \Phi_{(x_0,y_0)})^* g - \breve{g}\|_{C^{l,\beta}(B_2)}. \tag{6.2}$$

In $(\widetilde{\theta}, \rho)$ coordinates, we have

$$\Psi_{\widehat{p},r} \circ \Phi_{(x_0,y_0)}(x, y) = (rx_0 + ry_0 x, ry_0 y).$$

Let us abbreviate this composite map as $\zeta(x, y) = (rx_0 + ry_0 x, ry_0 y)$, so that

$$(\Psi_{\widehat{p},r} \circ \Phi_{(x_0,y_0)})^* g - \breve{g} = y^{-2} \zeta^*(\overline{g}_{ij} - \delta_{ij}) dx^i \, dx^j.$$

Since the $C^{l,\beta}(B_2)$ norm in (6.2) is just the norm of the components in (x,y)-coordinates, and y^{-2} is uniformly bounded on B_2 together with all its derivatives, it suffices to show that

$$\|\zeta^* f\|_{C^{l,\beta}(B_2)} \leq Cr\|f\|_{C^{l,\beta}_{(0)}}$$

for any function $f \in C^{l,\beta}_{(0)}(\Omega)$ that vanishes at $\widehat{p}$. Moreover, the $(\widetilde{\theta}, \rho)$ coordinates are uniformly $C^{l+1,\beta}$-equivalent to the original background coordinates (θ, ρ), because the coefficients B^α, the matrix (A^α_β), and its inverse are uniformly bounded. Thus the $C^{l,\beta}_{(0)}$ norm of f in $(\widetilde{\theta}, \rho)$ coordinates is uniformly bounded by the global norm $\|f\|_{C^{l,\beta}_{(0)}}$.

To bound the sup norm of $\zeta^* f$, we use the mean value theorem and the fact that $f(0,0) = f(\widehat{p}) = 0$ to estimate

$$|\zeta^* f(x,y)| = |f(rx_0 + ry_0 x, ry_0 y) - f(0,0)|$$
$$= |df_{(a_0,b_0)}(rx_0 + ry_0 x, ry_0 y)|$$
$$\leq Cr\|f\|_{C^{1,0}_{(0)}}$$

where (a_0, b_0) is some point on the line between $(0,0)$ and $(rx_0 + ry_0 x, ry_0 y)$. For any coordinate x^k ($k = 1, \ldots, n+1$), we have

$$|\partial_{x^k}(\zeta^* f)(x,y)| = |ry_0 \partial_{\theta^k} f(rx_0 + ry_0 x, ry_0 y)|$$
$$\leq r\|f\|_{C^{1,0}_{(0)}}.$$

The Hölder norm of the first derivatives is estimated as follows:

$$\frac{|\partial_{x^k}(\zeta^* f)(x,y) - \partial_{x^k}(\zeta^* f)(x',y')|}{|(x,y)-(x',y')|^\alpha}$$
$$= \frac{|ry_0 \partial_{\theta^k} f(rx_0 + ry_0 x, ry_0 y) - ry_0 \partial_{\theta^k} f(rx_0 + ry_0 x', ry_0 y')|}{|(x,y)-(x',y')|^\alpha}$$
$$\leq \frac{r\|f\|_{C^{1,\alpha}_{(0)}} |(rx_0 + ry_0 x, ry_0 y) - (rx_0 + ry_0 x', ry_0 y')|^\alpha}{|(x,y)-(x',y')|^\alpha}$$
$$\leq Cr^{1+\alpha} \|f\|_{C^{1,\alpha}_{(0)}}.$$

The general case now follows by induction on l, using the fact that $\partial_{x^k}(\zeta^* f) = ry_0 \Phi^*_{\widehat{p}}(\partial_{\theta^k} f)$. □

We need to explore how the weighted Sobolev and Hölder norms behave under boundary Möbius charts. The function y is not a defining function for hyperbolic space because it blows up at infinity; however, by patching together via a partition of unity, it is easy to construct a smooth defining function ρ_0 for hyperbolic space that is equal to y on Y_1. Then for any boundary Möbius chart $\Psi_{\widehat{p},r}$, it follows that $\Psi^*_{\widehat{p},r} \rho = y = \rho_0$ on Y_1 for r sufficiently small, so by the same reasoning that led to (3.4) and (3.5), the weighted norms have the following scaling behavior under boundary Möbius charts:

$$C^{-1} r^{-\delta} \|\Psi^*_{\widehat{p},r} u\|_{k,\alpha;Y_1} \leq \|u\|_{k,\alpha,\delta;Z_r(\widehat{p})} \leq C r^{-\delta} \|\Psi^*_{\widehat{p},r} u\|_{k,\alpha;Y_1}, \tag{6.3}$$

$$C^{-1} r^{-\delta} \|\Psi^*_{\widehat{p},r} u\|_{k,p;Y_1} \leq \|u\|_{k,p,\delta;Z_r(\widehat{p})} \leq C r^{-\delta} \|\Psi^*_{\widehat{p},r} u\|_{k,p;Y_1}. \tag{6.4}$$

Choose a specific smooth bump function $\psi \colon \mathbb{H} \to [0,1]$ that is equal to 1 on $A_{1/2}$ and supported in A_1. For any $\widehat{p} \in \partial M$ and any $r > 0$, let $(\widetilde{\theta}, \rho)$ be the coordinates on a neighborhood Ω of $\widehat{p}$ constructed above, and define $\psi_{\widehat{p},r} \in C^{l,\beta}(\Omega)$ by

$$\psi_{\widehat{p},r}(\widetilde{\theta}, \rho) = (\Psi^{-1}_{\widehat{p},r})^* \psi = \psi(\widetilde{\theta}/r, \rho/r).$$

Because the different choices of $(\widetilde{\theta}, \rho)$ coordinates are all uniformly bounded in $C^{l+1,\beta}_{(0)}(\Omega)$ with respect to each other, and $|d\rho|_g$, $|d\widetilde{\theta}^\alpha|_g$ are both in $C^{l,\beta}_1(M)$ with norms independent of $\widehat{p}$, it follows that the functions $\psi_{\widehat{p},r}$ are uniformly bounded in $C^{l,\beta}(\Omega)$, independently of $\widehat{p}$ and r.

By the same argument as in Lemma 2.2, there is a number N such that for any $r > 0$ we can choose (necessarily finitely many) points $\{\widehat{p}_1, \ldots, \widehat{p}_m\} \subset \partial M$ such that the sets $\{Z_{r/2}(\widehat{p}_i)\}$ cover $A_{r/2} = \{p \in M : \rho(p) < r/2\}$ and no more than N of the sets $\{Z_r(\widehat{p}_i)\}$ intersect nontrivially at any point. For any such covering, let $\Psi_i = \Psi_{\widehat{p}_i,r}$ and $\psi_i = \psi_{\widehat{p}_i,r}$. Let $\psi_0 \in C^\infty_c(M)$ be a smooth bump function that is supported in $M \smallsetminus A_{r/4}$ and equal to 1 on $M \smallsetminus A_{r/2}$, and define

$$\varphi_i = \frac{\psi_i}{\left(\sum_{j=0}^m \psi_i^2\right)^{1/2}}.$$

It follows that $\{\varphi_i^2\}$ is a partition of unity for $\overline{M}$ subordinate to the cover $\{M \smallsetminus A_{r/4}, Z_r(\widehat{p}_i)\}$. Moreover, at each point of $\overline{M}$, at least one of the functions ψ_i is equal to 1 at and at most N of them are nonzero, so the functions φ_i are still uniformly bounded in $C^{l,\beta}_{(0)}(\overline{M})$.

Let $\check{E}$ be the tensor bundle over hyperbolic space associated with the same $\mathrm{O}(n+1)$ or $\mathrm{SO}(n+1)$ representation as E, and let $\check{P}$ be the operator on hyperbolic space with the same local coordinate expression as P. For each boundary Möbius chart Ψ_i, let g_i be the metric $\Psi_i^* g$ defined on $Y_1 \subset \mathbb{H}$, and let $P_i \colon C^\infty(Y_1; E) \to C^\infty(Y_1; E)$ be the operator defined by

$$P_i u := \Psi_i^* P(\Psi_i^{-1*} u).$$

Then Lemma 6.1 implies that P_i is close to P in the following sense: For each $\delta \in \mathbb{R}$, $0 < \alpha < 1$, $1 < p < \infty$, and k such that $m \leq k \leq l$ and $m < k + \alpha \leq l + \beta$, there is a constant C (independent of r or i) such that for all $u \in C^{k,\alpha,\delta}(M; E)$,

$$\|P_i u - P u\|_{k-m,\alpha,\delta} \leq Cr \|u\|_{k,\alpha,\delta}, \tag{6.5}$$

and for all $u \in H^{k,p,\delta}(M; E)$,

$$\|P_i u - P u\|_{k-m,p,\delta} \leq Cr \|u\|_{k,p,\delta}. \tag{6.6}$$

Now assume that P satisfies the hypotheses of Theorem C. In particular, there is some constant C such that the L^2 estimate (1.4) holds on the complement of some compact set. Choosing $\hat{p} \in \partial M$ arbitrarily and r sufficiently small, (6.6) and (6.4) together imply that $\check{P}$ satisfies an analogous estimate (perhaps with a larger constant) for all smooth sections u of $\check{E}$ compactly supported in Y_1. But if $u \in C_c^\infty(\mathbb{H}; E)$ is arbitrary, there is a Möbius transformation that takes $\operatorname{supp} u$ into Y_1, so the same estimate holds globally on $\mathbb{H}$. Therefore, by the results of Chapter 5, $\check{P}$ is invertible on $C_\delta^{k,\alpha}(\mathbb{H}; E)$ for $|\delta - n/2| < R$, and on $H_\delta^{k,p}(\mathbb{H}; E)$ for $|\delta + n/p - n/2| < R$.

For any sufficiently small $r > 0$, define operators $Q_r, S_r, T_r \colon C_c^\infty(M; E) \to C_c^\infty(M; E)$ by

$$Q_r u = \sum_i \varphi_i (\Psi_i^{-1})^* \check{P}^{-1} \Psi_i^* (\varphi_i u),$$

$$S_r u = \sum_i \varphi_i (\Psi_i^{-1})^* \check{P}^{-1} (P_i - \check{P}) \Psi_i^* (\varphi_i u),$$

$$T_r u = \sum_i \varphi_i (\Psi_i^{-1})^* \check{P}^{-1} \Psi_i^* ([\varphi_i, P] u).$$

PROPOSITION 6.2. *Let $P \colon C^\infty(M; E) \to C^\infty(M; E)$ satisfy the hypotheses of Theorem C.*

(a) *If $|\delta + n/p - n/2| < R$ and $1 < p < \infty$, then Q_r, S_r, and T_r extend to bounded maps as follows:*

$$Q_r \colon H_\delta^{0,p}(M; E) \to H_\delta^{m,p}(M; E),$$
$$S_r \colon H_\delta^{m,p}(M; E) \to H_\delta^{m,p}(M; E),$$
$$T_r \colon H_\delta^{m-1,p}(M; E) \to H_{\delta_1}^{m,p}(M; E),$$

for any δ_1 such that $\delta \leq \delta_1 \leq \delta + 1$ and $|\delta_1 + n/p - n/2| < R$. Moreover, there exists $r_0 > 0$ such that if $u \in H_\delta^{m,p}(M; E)$ is supported in A_r for $0 < r < r_0$, then

$$Q_r P u = u + S_r u + T_r u \tag{6.7}$$

and

$$\|S_r u\|_{m,p,\delta} \leq Cr \|u\|_{m,p,\delta} \tag{6.8}$$

for some constant C independent of r and u.

(b) *If $|\delta - n/2| < R$, $0 < \alpha < 1$, and $m + \alpha \leq l + \beta$, then Q_r, S_r, and T_r extend to bounded maps as follows:*

$$Q_r \colon C^{0,\alpha}_\delta(M;E) \to C^{m,\alpha}_\delta(M;E),$$
$$S_r \colon C^{m,\alpha}_\delta(M;E) \to C^{m,\alpha}_\delta(M;E),$$
$$T_r \colon C^{m-1,\alpha}_\delta(M;E) \to C^{m,\alpha}_{\delta_1}(M;E),$$

for any δ_1 such that $\delta \leq \delta_1 \leq \delta + 1$ and $|\delta_1 - n/2| < R$. Moreover, there exists $r_0 > 0$ such that if $u \in C^{m,\alpha}_\delta(M;E)$ is supported in A_r for $0 < r < r_0$, then

$$Q_r P u = u + S_r u + T_r u \qquad (6.9)$$

and

$$\|S_r u\|_{m,\alpha,\delta} \leq Cr \|u\|_{m,\alpha,\delta} \qquad (6.10)$$

for some constant C independent of r and u.

PROOF. The fact that $Q_r P u = u + S_r u + T_r u$ in A_r is just a computation:

$$Q_r P u = \sum_i \varphi_i (\Psi_i^{-1})^* \check{P}^{-1} \Psi_i^* (\varphi_i P u)$$
$$= \sum_i \varphi_i (\Psi_i^{-1})^* \check{P}^{-1} \Psi_i^* (P(\varphi_i u)) + \sum_i \varphi_i (\Psi_i^{-1})^* \check{P}^{-1} \Psi_i^* ([\varphi_i, P] u)$$
$$= \sum_i \varphi_i (\Psi_i^{-1})^* \check{P}^{-1} P_i \Psi_i^* (\varphi_i u) + T_r u$$
$$= \sum_i \varphi_i (\Psi_i^{-1})^* \check{P}^{-1} \check{P} \Psi_i^* (\varphi_i u)$$
$$\quad + \sum_i \varphi_i (\Psi_i^{-1})^* \check{P}^{-1} (P_i - \check{P}) \Psi_i^* (\varphi_i u) + T_r u$$
$$= u + S_r u + T_r u.$$

To check the mapping properties of S_r, we begin by observing that the fact that the functions φ_i are uniformly bounded in $C^{l,\beta}_{(0)}(\overline{M}) \subset C^{l,\beta}(M)$ implies by Lemma 3.6(a) that multiplication by φ_i is a bounded map from $H^{j,p}_\delta(Z_r(\widehat{p}_i);E)$ to itself for each i and all $0 \leq j \leq l$, with norm bounded independently of i and r. The fact that S_r maps $H^{m,p}_\delta(M;E)$ to itself then follows from Proposition 5.8 and (6.4), because the factors of r^δ and $r^{-\delta}$ introduced by Ψ_i^* and its inverse cancel each other. Moreover, (6.5) implies (6.8) whenever u is supported in A_r, for some constant C independent of r and u.

The mapping properties of T_r will follow from a similar argument once we show that the commutator $[\varphi_i, P]$ maps $H^{m-1,p}_\delta(M;E)$ to $H^{m,p}_{\delta_1}(M;E)$. Observe that each term in the coordinate expression for $[\varphi_i, P]u$ is a product of four factors: a constant, a pth covariant derivative of u, a qth covariant derivative of φ_i, and a polynomial in the components of g, $(\det g)^{-1/2}$, and their derivatives up through order r, with $p + q + r \leq m$ and $q \geq 1$. Since φ_i is uniformly bounded in $C^{l,\beta}(M)$, the result follows. The argument for the Hölder case is identical. □

COROLLARY 6.3. *Let $P \colon C^\infty(M;E) \to C^\infty(M;E)$ satisfy the hypotheses of Theorem C.*

(a) If $|\delta + n/p - n/2| < R$, $|\delta_1 + n/p - n/2| < R$, $\delta \leq \delta_1 \leq \delta + 1$, and $1 < p < \infty$, then there exist $r > 0$ and bounded operators
$$\widetilde{Q} \colon H^{0,p}_\delta(M;E) \to H^{m,p}_\delta(M;E),$$
$$\widetilde{T} \colon H^{m-1,p}_\delta(M;E) \to H^{m,p}_{\delta_1}(M;E)$$
such that
$$\widetilde{Q}Pu = u + \widetilde{T}u$$
whenever $u \in H^{m,p}_\delta(M;E)$ is supported in A_r.

(b) If $|\delta - n/2| < R$, $|\delta_1 - n/2| < R$, $\delta \leq \delta_1 \leq \delta + 1$, $0 < \alpha < 1$, and $m + \alpha \leq l + \beta$, then there exist $r > 0$ and bounded operators
$$\widetilde{Q} \colon C^{0,\alpha}_\delta(M;E) \to C^{m,\alpha}_\delta(M;E),$$
$$\widetilde{T} \colon C^{m-1,\alpha}_\delta(M;E) \to C^{m,\alpha}_{\delta_1}(M;E)$$
such that
$$\widetilde{Q}Pu = u + \widetilde{T}u \tag{6.11}$$
whenever $u \in C^{m,\alpha}_\delta(M;E)$ is supported in A_r.

PROOF. Just choose r small enough that (6.8) holds with $Cr < 1/2$. It then follows that $\mathrm{Id} + S_r \colon H^{m,p}_\delta(M;E) \to H^{m,p}_\delta(M;E)$ has a bounded inverse. We just set
$$\widetilde{Q} = (\mathrm{Id} + S_r)^{-1} \circ Q_r,$$
$$\widetilde{T} = (\mathrm{Id} + S_r)^{-1} \circ T_r,$$
and then (6.11) follows immediately from (6.7). Once again, the argument for the Hölder case is identical. $\square$

Our first application of this parametrix construction is a significant strengthening of Lemma 4.8, giving improved decay for solutions to $Pu = f$ when u and f are in appropriate spaces. We begin with a special case.

LEMMA 6.4. *Assume P satisfies the hypotheses of Theorem C.*

(a) *Suppose that $1 < p < \infty$, $m \leq k \leq l$, $|\delta + n/p - n/2| < R$, and $|\delta' + n/p - n/2| < R$. If $u \in H^{0,p}_\delta(M;E)$ and $Pu \in H^{k-m,p}_{\delta'}(M;E)$, then $u \in H^{k,p}_{\delta'}(M;E)$.*
(b) *Suppose that $0 < \alpha < 1$, $m < k + \alpha \leq l + \beta$, $|\delta - n/2| < R$, and $|\delta' - n/2| < R$. If $u \in C^{0,0}_\delta(M;E)$ and $Pu \in C^{k-m,\alpha}_{\delta'}(M;E)$, then $u \in C^{k,\alpha}_{\delta'}(M;E)$.*

PROOF. If $\delta' \leq \delta$, the result is a trivial consequence of Lemmas 4.8 and 3.6(b), so assume $\delta' > \delta$. Consider part (a). By Lemma 4.8, it suffices to show that $u \in H^{0,p}_{\delta'}(M;E)$. For any small $r > 0$, by means of a bump function we can write $u = u_0 + u_\infty$, where $\mathrm{supp}\, u_0$ is compact and $\mathrm{supp}\, u_\infty \subset A_r$. Local elliptic regularity gives $u_0 \in H^{k,p}_{\delta'}(M;E)$. Since Pu_∞ agrees with Pu off of a compact set, $Pu_\infty \in H^{0,p}_{\delta'}(M;E)$, so by Corollary 6.3, if r is small enough,
$$u_\infty = \widetilde{Q}Pu_\infty - \widetilde{T}u_\infty$$
$$\in H^{m,p}_{\delta'}(M;E) + H^{1,p}_{\delta_1}(M;E)$$
$$\subset H^{0,p}_{\delta_1}(M;E),$$

where $\delta_1 = \min(\delta', \delta+1)$. Iterating this argument finitely many times, we conclude that $u_\infty \in H_{\delta'}^{0,p}(M;E)$. By Lemma 3.6, this implies that $u \in H_{\delta'}^{k,p}(M;E)$ as claimed. The argument for the Hölder case is the same. □

PROPOSITION 6.5. *Suppose P satisfies the hypotheses of Theorem C, and u is either in $H_{\delta_0}^{0,p_0}(M;E)$ for some $|\delta_0 + n/p_0 - n/2| < R$ and $1 < p_0 < \infty$, or in $C_{\delta_0}^{0,0}(M;E)$ for some $|\delta_0 - n/2| < R$.*

(a) *If $Pu \in H_\delta^{k-m,p}(M;E)$ for $|\delta + n/p - n/2| < R$, $1 < p < \infty$, and $m \leq k \leq l$, then $u \in H_\delta^{k,p}(M;E)$.*
(b) *If $Pu \in C_\delta^{k-m,\alpha}(M;E)$ for $|\delta - n/2| < R$, $0 < \alpha < 1$, and $m < k + \alpha \leq l + \beta$, then $u \in C_\delta^{k,\alpha}(M;E)$.*

PROOF. If $u \in C_{\delta_0}^{0,0}(M;E)$ with $|\delta_0 - n/2| < R$, then $u \in H_\delta^{0,p}(M;E)$ whenever $\delta + n/p < \delta_0$ by Lemma 3.6. Since such δ and p can be chosen that also satisfy $|\delta + n/p - n/2| < R$, it suffices to prove the proposition under the hypothesis that $u \in H_{\delta_0}^{0,p_0}(M;E)$ with $|\delta_0 + n/p_0 - n/2| < R$. For the rest of the proof, we assume this.

First we treat case (a). Assume that $Pu \in H_\delta^{k-m,p}(M;E)$ with $|\delta + n/p - n/2| < R$, and let $\mathscr{P}$ be the following set:

$$\mathscr{P} = \{p' \in (1,\infty) : u \in H_{\delta'}^{0,p'}(M;E) \text{ for some } \delta' \text{ with } |\delta' + n/p' - n/2| < R\}.$$

Clearly $p_0 \in \mathscr{P}$ by hypothesis. We will show that $p \in \mathscr{P}$. It will then follow from Lemma 6.4 that $u \in H_\delta^{k,p}(M;E)$, which will prove case (a).

CLAIM 1: If $p_1 \in \mathscr{P}$, then $(1, p_1] \subset \mathscr{P}$. To prove this, assume $p_1 \in \mathscr{P}$ and $1 < p' < p_1$. The fact that $p_1 \in \mathscr{P}$ means that there is some δ_1 with $|\delta_1 + n/p_1 - n/2| < R$ such that $u \in H_{\delta_1}^{0,p_1}(M;E)$. By Lemma 3.6, $u \in H_{\delta'}^{0,p'}(M;E)$ for any δ' such that $\delta_1 + n/p_1 > \delta' + n/p'$. Choosing δ' so that $\delta' + n/p'$ is sufficiently close to $\delta_1 + n/p_1$, we can ensure that $|\delta' + n/p' - n/2| < R$. This implies that $p' \in \mathscr{P}$ as desired.

CLAIM 2: If $p_1 \in \mathscr{P}$ and p_2 satisfies $p_1 < p_2 \leq p$ and

$$\frac{p_2}{p_1} \leq \min\left(\frac{n+2}{n+1}, 1 + \frac{\varepsilon}{2n}\right), \tag{6.12}$$

where

$$\varepsilon = \delta + \frac{n}{p} - \frac{n}{2} + R > 0,$$

then $p_2 \in \mathscr{P}$. The assumption that $p_1 \in \mathscr{P}$ means that $u \in H_{\delta_1}^{0,p_1}(M;E)$ for some δ_1 with $|\delta_1 + n/p_1 - n/2| < R$. Choose δ' satisfying

$$\delta + \frac{n}{p} > \delta' + \frac{n}{p_1} > \delta + \frac{n}{p} - \frac{\varepsilon}{2}. \tag{6.13}$$

By virtue of the first inequality above, Lemma 3.6 implies that

$$Pu \in H_\delta^{k-m,p}(M;E) \subset H_{\delta'}^{k-m,p_1}(M;E). \tag{6.14}$$

Since the two inequalities of (6.13) guarantee that $|\delta' + n/p_1 - n/2| < R$, Lemma 6.4 shows that $u \in H^{k,p_1}_{\delta'}(M;E)$. Our restriction on p_2 guarantees that

$$\frac{n+1}{p_1} \leq \frac{n+2}{p_2}$$
$$= \frac{n+1}{p_2} + \frac{1}{p_2}$$
$$\leq \frac{n+1}{p_2} + k,$$

so Lemma 3.6(c) implies that $u \in H^{0,p_2}_{\delta'}(M;E)$. Now (6.12) implies that

$$\frac{n}{p_1} - \frac{n}{p_2} = \frac{n}{p_2}\left(\frac{p_2}{p_1} - 1\right)$$
$$\leq \frac{n}{p_2}\left(\frac{\varepsilon}{2n}\right)$$
$$< \frac{\varepsilon}{2}.$$

Therefore, using (6.13), we obtain

$$\delta' + \frac{n}{p_2} - \frac{n}{2} = \left(\delta' + \frac{n}{p_1} - \frac{n}{2}\right) - \left(\frac{n}{p_1} - \frac{n}{p_2}\right)$$
$$> \left(\delta + \frac{n}{p} - \frac{n}{2} - \frac{\varepsilon}{2}\right) - \left(\frac{\varepsilon}{2}\right)$$
$$= -R,$$
$$\delta' + \frac{n}{p_2} - \frac{n}{2} < \delta + \frac{n}{p} - \frac{n}{2}$$
$$< R,$$

which proves that $p_2 \in \mathscr{P}$ as claimed.

CLAIM 3: $p \in \mathscr{P}$. If $p \leq p_0$, this follows immediately from Claim 1 together with the obvious fact that $p_0 \in \mathscr{P}$. Otherwise, just iterate Claim 2, starting with $p_0 \in \mathscr{P}$. After finitely many iterations, we can conclude that $p \in \mathscr{P}$.

Finally we turn to case (b). Suppose $Pu \in C^{k-m,\alpha}_{\delta}(M;E)$ with $|\delta - n/2| < R$, and choose p' large and δ' close to δ satisfying

$$\delta > \delta' + \frac{n}{p'}, \qquad (6.15)$$

Then by Lemma 3.6, $Pu \in H^{k-m,p'}_{\delta'}(M;E)$. If we choose $\delta' + n/p'$ sufficiently close to δ, we have $|\delta' + n/p' - n/2| < R$, and thus $u \in H^{k,p'}_{\delta'}(M;E)$ by part (a) above. If p is also chosen large enough that $(n+1)/p' \leq k - \alpha$, the Sobolev embedding theorem (Theorem 3.6(c)) implies $u \in C^{0,\alpha}_{\delta'}(M;E)$, and if δ' is sufficiently close to δ we will have $|\delta' - n/2| < R$. Then Lemma 6.4 implies $u \in C^{k,\alpha}_{\delta}(M;E)$. □

Now suppose $P \colon C^\infty(M;E) \to C^\infty(M;E)$ satisfies the hypotheses of Theorem C. By Lemma 4.10, estimate (1.4) implies that $P \colon H^{m,2}(M;E) \to H^{0,2}(M;E)$ is Fredholm. Let $Z = \operatorname{Ker} P \cap L^2(M;E)$, which is equal to $\operatorname{Ker} P \cap H^{m,2}(M;E)$ by

Lemma 4.8. Then Z is finite-dimensional, and Proposition 6.5 shows that $Z \subset H_\delta^{k,p}(M;E)$ whenever $1 < p < \infty$, $m \le k \le l$, and $|\delta + n/p - n/2| < R$. In fact,

$$Z = \operatorname{Ker} P \colon H_\delta^{k,p}(M;E) \to H_\delta^{k-m,p}(M;E), \tag{6.16}$$

because any $u \in \operatorname{Ker} P \cap H_\delta^{k,p}(M;E)$ is also in $L^2(M;E)$ by Proposition 6.5. Similarly, if $0 < \alpha < 1$, $m < k + \alpha \le l + \beta$, and $|\delta - n/2| < R$, then $Z = \operatorname{Ker} P \colon C_\delta^{k,\alpha}(M;E) \to C_\delta^{k-m,\alpha}(M;E)$.

Because of these observations, whenever $1 < p < \infty$, $m \le k \le l$, and $|\delta + n/p - n/2| < R$, we can define a subspace $Y_\delta^{k,p} \subset H_\delta^{k,p}(M;E)$ by

$$Y_\delta^{k,p} = \{u \in H_\delta^{k,p}(M;E) : (u,v) = 0 \text{ for all } v \in Z\},$$

where (u,v) represents the standard L^2 pairing. Since $Z \subset H_{-\delta}^{0,p^*}(M;E) = (H_\delta^{0,p}(M;E))^* \subset (H_\delta^{k,p}(M;E))^*$, it follows that $Y_\delta^{k,p}$ is a well-defined closed subspace of $H_\delta^{k,p}(M;E)$. Similarly, if $0 < \alpha < 1$, $m < k+\alpha \le l+\beta$, and $|\delta - n/2| < R$, we can define

$$Y_\delta^{k,\alpha} = \{u \in C_\delta^{k,\alpha}(M;E) : (u,v) = 0 \text{ for all } v \in Z\},$$

since $C_\delta^{k,\alpha}(M;E) \subset H_{\delta'}^{0,p}(M;E)$ for $\delta > \delta' + n/p > n/2 - R$ implies $Z \subset H_{-\delta'}^{0,p^*}(M;E) = (H_{\delta'}^{0,p}(M;E))^* \subset (C_\delta^{k,\alpha}(M;E))^*$.

The next result is the main structure theorem for operators satisfying the hypotheses of Theorem C.

THEOREM 6.6. *Suppose $P \colon C^\infty(M;E) \to C^\infty(M;E)$ satisfies the hypotheses of Theorem C.*

(a) *If $1 < p < \infty$, $0 \le k \le l$, and $|\delta + n/p - n/2| < R$, there exist bounded operators $G, H \colon H_\delta^{k,p}(M;E) \to H_\delta^{k,p}(M;E)$ such that $G(H_\delta^{k-m,p}(M;E)) \subset H_\delta^{k,p}(M;E)$ for $k \ge m$, and*

$$Y_\delta^{k,p} = \operatorname{Ker} H, \tag{6.17}$$
$$Z = \operatorname{Im} H, \tag{6.18}$$
$$u = GPu + Hu \text{ for } u \in H_\delta^{k,p}(M;E),\ m \le k \le l, \tag{6.19}$$
$$u = PGu + Hu \text{ for } u \in H_\delta^{k,p}(M;E),\ 0 \le k \le l. \tag{6.20}$$

(b) *If $0 < \alpha < 1$, $0 < k+\alpha \le l+\beta$, and $|\delta - n/2| < R$, there exist bounded operators $G, H \colon C_\delta^{k,\alpha}(M;E) \to C_\delta^{k,\alpha}(M;E)$ such that $G(C_\delta^{k-m,\alpha}(M;E)) \subset C_\delta^{k,\alpha}(M;E)$ for $k \ge m$, and*

$$Y_\delta^{k,\alpha} = \operatorname{Ker} H, \tag{6.21}$$
$$Z = \operatorname{Im} H, \tag{6.22}$$
$$u = GPu + Hu \text{ for } u \in C_\delta^{k,\alpha}(M;E),\ m \le k \le l, \tag{6.23}$$
$$u = PGu + Hu \text{ for } u \in C_\delta^{k,\alpha}(M;E),\ 0 \le k \le l. \tag{6.24}$$

PROOF. We begin with the Sobolev case, part (a). First consider the special case $p = 2$, $k = 0$, and $\delta = 0$ (in which case this is basically the standard construction of a partial inverse for a Fredholm operator on L^2). As noted above, the assumption of an L^2 estimate (1.4) near the boundary implies that $P \colon H_0^{m,2}(M;E) \to L^2(M;E)$ is Fredholm by Lemma 4.10.

By definition, $Y_0^{0,2} \subset L^2(M;E)$ is precisely the orthogonal complement of $Z =$ Ker P in $L^2(M;E)$, so we have an orthogonal direct sum decomposition $L^2(M;E) = Z \oplus Y_0^{0,2}$. Since P is formally self-adjoint, any $u \in H^{m,2}(M;E)$ satisfies $(Pu, v) = (u, Pv) = 0$ for all $v \in Z$, so $P(H^{m,2}(M;E)) \subset Y_0^{0,2}$. On the other hand, if $v \in L^2(M;E)$ is orthogonal to $P(H^{m,2}(M;E))$, then for any smooth, compactly supported section u of E we have $(v, Pu) = 0$, so v is a distributional solution to $Pv = 0$, which means $v \in Z$. This shows that $(P(H^{m,2}(M;E)))^\perp = Z$, and since $P(H^{m,2}(M;E))$ is closed in $L^2(M;E)$ we have $P(H^{m,2}(M;E)) = Y_0^{0,2}$.

Now $P \colon Y_0^{m,2} \to Y_0^{0,2}$ is bijective and bounded, so by the open mapping theorem it has a bounded inverse $(P|_{Y_0^{m,2}})^{-1} \colon Y_0^{0,2} \to Y_0^{m,2}$. Define $G \colon L^2(M;E) \to L^2(M;E)$ by

$$Gu = \begin{cases} (P|_{Y_0^{m,2}})^{-1} u & u \in Y_0^{0,2}, \\ 0 & u \in Z, \end{cases}$$

and define $H \colon L^2(M;E) \to L^2(M;E)$ to be the orthogonal projection onto Z. Then (6.17) and (6.18) are immediate from the definition of H, and (6.19) (for $u \in H^{m,2}(M;E)$) and (6.20) (for all u) follow by considering $u \in Y_0^{0,2}$ and $u \in Z$ separately.

Next consider the case of arbitrary δ satisfying $|\delta| < R$, still with $p = 2$ and $k = 0$. If $\delta > 0$ and $u \in H_\delta^{0,2}(M;E) \subset L^2(M;E)$, then $PGu = u - Hu \in H_\delta^{0,2}(M;E)$, so $Gu \in H_\delta^{m,2}(M;E)$ by Lemma 6.4. Thus the restriction of G to $H_\delta^{0,2}(M;E)$ takes its values in $H_\delta^{m,2}(M;E)$, as does H by (6.16). In this case, (6.17)–(6.20) are satisfied because they are already satisfied on the bigger space $L^2(M;E)$ (or $H^{m,2}(M;E)$ in case of (6.19)), and $Y_\delta^{0,2} = Y_0^{0,2} \cap H_\delta^{0,2}(M;E)$.

On the other hand, if $\delta < 0$, we can use the fact that $H_\delta^{0,2}(M;E) = (H_{-\delta}^{0,2}(M;E))^*$ to extend the definition of G and H to $H_\delta^{0,2}(M;E)$ by duality: For any $u \in H_\delta^{0,2}(M;E)$, let Gu and Hu be the elements of $H_\delta^{0,2}(M;E)$ defined uniquely by

$$(Gu, v) = (u, Gv), \tag{6.25}$$
$$(Hu, v) = (u, Hv) \tag{6.26}$$

for all $v \in H_{-\delta}^{0,2}(M;E)$. In other words, $G, H \colon H_\delta^{0,2}(M;E) \to H_\delta^{0,2}(M;E)$ are defined to be the dual maps of $G, H \colon H_{-\delta}^{0,2}(M;E) \to H_{-\delta}^{0,2}(M;E)$. Since H and G are self-adjoint on $L^2(M;E)$ (H because it is an orthogonal projection, and G because P is self-adjoint as an unbounded operator), these are indeed extensions of the original maps G and H.

To see that these extended operators satisfy (6.17)–(6.20), we observe that (6.26) implies that $Hu = 0$ for $u \in H_\delta^{0,2}(M;E)$ exactly when $(u, v) = 0$ for all v in the image of $H \colon H_{-\delta}^{0,2}(M;E) \to H_{-\delta}^{0,2}(M;E)$; since this image is exactly Z, it follows that Ker $H = Y_\delta^{0,2}$, which is (6.17). Since the restriction of H to $Z \subset L^2(M;E) \subset H_\delta^{0,2}(M;E)$ is the identity, it follows that $Z \subset \mathrm{Im}\, H$. On the other hand, for any $u \in H_\delta^{0,2}(M;E)$, we have $(Hu, Pv) = (u, HPv) = 0$ for all $v \in C_c^\infty(M;E) \subset H_{-\delta}^{0,2}(M;E)$, which means that Hu is a weak solution to $P(Hu) = 0$. Thus $\mathrm{Im}\, H \subset \mathrm{Ker}\, P = Z$, which proves (6.18). Equations (6.19) and (6.20) then follow easily from our definitions by duality.

Next we generalize to $1 < p < \infty$ and $|\delta + n/p - n/2| < R$, still with $k = 0$. If $p > 2$, we can choose δ' such that $|\delta'| < R$ and $\delta + n/p > \delta' + n/2$, so that $H_\delta^{0,p}(M;E) \subset H_{\delta'}^{0,2}(M;E)$. Arguing as above, we see that the restrictions of G and H map $H_\delta^{0,p}(M;E)$ to $H_\delta^{m,p}(M;E)$. On the other hand, for $p < 2$, we can extend G and H to maps from $H_\delta^{0,p}(M;E) = (H_{-\delta}^{0,p^*}(M;E))^*$ to itself by duality as above. In both cases (6.17)–(6.20) are satisfied, by restriction or duality as appropriate.

Now consider the general case of $H_\delta^{k,p}(M;E)$ with $0 \le k \le l$, $1 < p < \infty$, and $|\delta + n/p - n/2| < R$. Since $Z \subset H_\delta^{k,p}(M;E)$, it is clear that H restricts to a map of $H_\delta^{k,p}(M;E)$ to itself. If $u \in H_\delta^{k,p}(M;E) \subset H_\delta^{0,p}(M;E)$ for $0 \le k \le l-m$, observe as above that $PGu = u - Hu \in H_\delta^{k,p}(M;E)$, so $Gu \in H_\delta^{k+m,p}(M;E)$ by Lemma 4.8. For $l - m \le k \le l$, we have $G(H_\delta^{k,p}(M;E)) \subset G(H_\delta^{l-m,p}(M;E)) \subset H_\delta^{l,p}(M;E) \subset H_\delta^{k,p}(M;E)$. Thus in each case G and H restrict to maps from $H_\delta^{k,p}(M;E)$ to itself, and properties (6.17)–(6.20) are satisfied by restriction.

Finally, consider case (b), and assume that $0 < \alpha < 1$, $0 < k + \alpha \le l + \beta$, and $|\delta - n/2| < R$. We can choose $\delta' \in \mathbb{R}$ satisfying $|\delta'| < R$ and $\delta > \delta' + n/2$, so that $C_\delta^{k,\alpha}(M;E) \subset H_{\delta'}^{k,2}(M;E)$ and the results of part (a) apply to $H_{\delta'}^{k,2}(M;E)$. Then the restrictions of G and H map $C_\delta^{k,\alpha}(M;E)$ to itself by the same argument as above, and properties (6.21)–(6.24) are automatically satisfied by restriction. □

The next construction will be useful in proving that P is not Fredholm outside the expected range of weights.

LEMMA 6.7. *Suppose $P\colon C^\infty(M;E) \to C^\infty(M;E)$ satisfies the hypotheses of Theorem C, where E is a bundle of tensors of weight r. Let s_0 be a characteristic exponent of P, and let $\delta_0 = \operatorname{Re} s_0 + r$. Given any compact subset $K \subset M$, there is an infinite-dimensional subspace $W \subset C_{\delta_0}^{l,\beta}(M;E)$ such that every nonzero $w \in W$ has the following properties.*

(a) $\operatorname{supp} w \subset M \smallsetminus K$.
(b) $Pw \in C_{\delta_0+1}^{0,0}(M;E)$.
(c) *If $1 < p < \infty$ and $\delta \ge \delta_0 - n/p$, then $w \notin H_\delta^{0,p}(M;E)$.*

PROOF. Let $\widehat{p} \in \partial M$ be arbitrary, and let V be any neighborhood of $\widehat{p}$ in $\overline{M}$. Since the characteristic exponents are constant on ∂M by Lemma 4.3, there is a tensor $\overline{w}_{\widehat{p}} \in E_{\widehat{p}}$ such that $I_{s_0}(P)\overline{w}_{\widehat{p}} = 0$. We can extend $\overline{w}_{\widehat{p}}$ to a $C^{l,\beta}$ tensor field $\overline{w}$ on a neighborhood of $\widehat{p}$ in ∂M, still satisfying $I_{s_0}(P)\overline{w} = 0$, as follows. Shrinking V if necessary, we can choose background coordinates on V, and for each $\widehat{q} \in V \cap \partial M$, let $A(\widehat{q})$ be the matrix of $I_{s_0}(P)\colon E_{\widehat{q}} \to E_{\widehat{q}}$. By Lemma 4.2, the matrix entries of $A(\widehat{q})$ are $C^{l,\beta}$ functions of $\widehat{q}$. If γ is any smooth, positively-oriented closed curve in $\mathbb{C}$ whose interior contains 0 but no other eigenvalues of $A(\widehat{p})$, then the projection onto the kernel of $A(\widehat{q})$ can be written as $-1/(2\pi i) \int_\gamma (A(\widehat{q}) - z\operatorname{Id})^{-1} dz$. (Here we use the fact that the eigenvalues of $A(\widehat{q})$ and their multiplicities are independent of $\widehat{q}$.) We define $\overline{w}$ to be the tensor field on $V \cap \partial M$ whose coordinate expression is

$$\overline{w}_{\widehat{q}} = \frac{-1}{2\pi i} \int_\gamma (A(\widehat{q}) - z\operatorname{Id})^{-1} \overline{w}_{\widehat{p}} dz,$$

which is a $C^{l,\beta}$ tensor field along ∂M satisfying $I_{s_0}(P)\overline{w} = 0$. If we extend $\overline{w}$ arbitrarily to a $C^{l,\beta}$ tensor field on V, and let $w = \rho^{s_0} \varphi \overline{w}$ where φ is any smooth

cutoff function that is positive at $\widehat{p}$ and supported in V, then $w \in C^{l,\beta}_\delta(M;E)$, and
$$|Pw|_{\overline{g}} = O(\rho^{\operatorname{Re} s_0+1}),$$
which implies
$$|Pw|_g = O(\rho^{\delta_0+1}), \tag{6.27}$$
so $Pw \in C^{0,0}_{\delta_0+1}(M;E)$. On the other hand, $w \notin H^{0,p}_\delta(M;E)$ for $\delta \geq \delta_0 - n/p$ by Lemma 3.2.

Now choose countably many points $\widehat{p}_i \in \partial M$ and disjoint neighborhoods V_i of $\widehat{p}_i$. For each i we can construct w_i as above with support in V_i, so the space spanned by $\{w_i\}$ is clearly infinite-dimensional. □

PROOF OF PROPOSITION B. It follows immediately from Lemma 4.10 that P is Fredholm as a map from $H^{m,2}(M;E)$ to $H^{0,2}(M;E)$ if and only if P satisfies an estimate of the form (1.4). By Lemma 4.8, the kernel and range of P as an unbounded operator on $L^2(M;E)$ are the same as those of $P\colon H^{m,2}(M;E) \to H^{0,2}(M;E)$. The proposition follows. □

Finally, we are in a position to prove our main Fredholm theorem, Theorem C from the Introduction.

PROOF OF THEOREM C. We will prove parts (b) and (c) together. The proof of sufficiency is identical for the Sobolev and Hölder cases, so we do only the Sobolev case.

Suppose $1 < p < \infty$, $m \leq k \leq l$, and $|\delta + n/p - n/2| < R$, and let $G, H\colon H^{k,p}_\delta(M;E) \to H^{k,p}_\delta(M;E)$ be as in Theorem 6.6. We have already remarked that the kernel of $P\colon H^{k,p}_\delta(M;E) \to H^{k-m,p}_\delta(M;E)$ is equal to Z, which is finite-dimensional, and in fact is the same as the L^2 kernel. We will show that the range of P is closed by showing that it is equal to $Y^{k-m,p}_\delta$. If $f = Pu \in P(H^{k,p}_\delta(M;E))$, then clearly $(f,v) = (Pu,v) = (u,Pv) = 0$ for all $v \in Z$, so $f \in Y^{k-m,p}_\delta$. On the other hand, if $f \in Y^{k-m,p}_\delta$, then $f = PGf + Hf = PGf$ by (6.20) and (6.17), so $f \in P(H^{k,p}_\delta(M;E))$. Since every $f \in H^{k-m,p}_\delta(M;E)$ can be written $f = PGf + Hf$, where $PGf \in P(H^{k,p}_\delta(M;E)) = Y^{k-m,p}_\delta$ and $Hf \in Z$, it follows that $H^{k-m,p}_\delta(M;E) = Y^{k-m,p}_\delta \oplus Z$. Therefore,
$$\frac{H^{k-m,p}_\delta(M;E)}{P(H^{k,p}_\delta(M;E))} = \frac{Y^{k-m,p}_\delta \oplus Z}{Y^{k-m,p}_\delta} \cong Z,$$
which is finite-dimensional. This also shows that the cokernel and kernel of P have the same dimension, so P has index zero.

Next we will prove the necessity of the stated conditions on δ. In fact, we will show that P has infinite-dimensional kernel when δ is strictly below the Fredholm range, and infinite-dimensional cokernel when δ is strictly above; in the borderline case $C^{k,\alpha}_{n/2-R}$, we will also show that P has infinite-dimensional kernel, and in all other borderline cases, we will show that it fails to have closed range.

First we address the Hölder case below the Fredholm range. Assume $0 < \alpha < 1$, $m < k+\alpha \leq l+\beta$, and $\delta \leq n/2 - R$, and consider $P\colon C^{k,\alpha}_\delta(M;E) \to C^{k-m,\alpha}_\delta(M;E)$. We will prove that P is not Fredholm in this case by showing that it has an infinite-dimensional kernel.

The definition of the indicial radius R and the symmetry of the characteristic exponents about $\operatorname{Re} s = n/2 - r$ imply that P has a characteristic exponent s_0 with

$\mathrm{Re}\, s_0 = n/2 - r - R$. Let W be the subspace of $C^{l,\beta}_{n/2-R}(M;E)$ given by Lemma 6.7 for this characteristic exponent. (The compact set K is irrelevant in this case.) For any $p > n$, $P(W) \subset C^{0,0}_{n/2-R+1}(M;E) \subset H^{0,p}_{n/2-R}(M;E)$ by Lemma 3.6. If we choose p large enough, then $\delta = n/2 - R$ will satisfy $|\delta + n/p - n/2| = |n/p - R| < R$, so there exist operators $G, H\colon H^{0,p}_{n/2-R}(M;E) \to H^{0,p}_{n/2-R}(M;E)$ satisfying (6.17)–(6.20). Let $W_0 \subset W$ be the linear subspace defined by

$$W_0 = \{w \in W : HPw = 0\}. \tag{6.28}$$

Because H takes its values in the finite-dimensional space Z, the space W_0 is also infinite-dimensional. Note that for $w \in W_0$, $Pw \in H^{0,p}_{n/2-R}(M;E)$ implies $GPw \in H^{m,p}_{n/2-R}(M;E) \subset C^{0,0}_{n/2-R}(M;E)$ by Lemma 3.6(c). Define $X\colon W_0 \to C^{0,0}_{n/2-R}(M;E)$ by $Xw = w - GPw$. It follows from (6.20) that

$$PXw = Pw - PGPw = HPw = 0 \quad \text{for all } w \in W_0.$$

Therefore,

$$X(W_0) \subset \mathrm{Ker}\, P \cap C^{0,0}_{n/2-R}(M;E) \subset C^{l,\beta}_{n/2-R}(M;E)$$

by Lemma 4.8. Moreover, X is injective because $Xw = 0$ implies $w = GPw \in H^{0,p}_{n/2-R}(M;E)$, which implies that $w = 0$ by assertion (c) of Lemma 6.7. Thus we have shown that $X(W_0)$ is an infinite-dimensional subspace of $\mathrm{Ker}\, P \cap C^{l,\beta}_{n/2-R}(M;E)$. Since $C^{l,\beta}_{n/2-R}(M;E) \subset C^{k,\alpha}_\delta(M;E)$ whenever $\delta \leq n/2 - R$ and $m < k + \alpha \leq l + \beta$, it follows that P has infinite-dimensional kernel on $C^{k,\alpha}_\delta(M;E)$ in all such cases.

Next consider the Sobolev case below the Fredholm range. When $1 < p < \infty$, $m \leq k \leq l$, and $\delta < n/2 - n/p - R$, we have $C^{l,\beta}_{n/2-R}(M;E) \subset H^{k,p}_\delta(M;E)$ by Lemma 3.6, so P has infinite-dimensional kernel in $H^{k,p}_\delta(M;E)$ as well.

Now we consider the exponents strictly above the Fredholm range, beginning with the Sobolev case. Suppose $1 < p < \infty$, $m < k + \alpha \leq l + \beta$, and $\delta > n/2 - n/p + R$. Recall that the dual space to $H^{0,p}_\delta$ is $H^{0,p^*}_{-\delta}$ (where p^* is the conjugate exponent, $1/p + 1/p^* = 1$), acting by way of the standard L^2 pairing. Since $-\delta < n/2 - n/p^* - R$, the argument above shows that $P^* = P\colon H^{m,p^*}_{-\delta}(M;E) \to H^{0,p^*}_{-\delta}(M;E)$ has infinite-dimensional kernel. Each element v of the infinite-dimensional space $\mathrm{Ker}\, P \cap H^{m,p^*}_{-\delta}(M;E)$ thus defines a continuous linear functional on $H^{k-m,p}_\delta(M;E)$ by $u \mapsto (u,v)$, and each such linear functional annihilates $P(H^{k,p}_\delta(M;E))$ by Lemma 4.7. It follows that the range of P has infinite codimension in $H^{k-m,p}_\delta(M;E)$.

For the Hölder case, suppose $0 < \alpha < 1$, $m < k + \alpha \leq l + \beta$, and $\delta > n/2 + R$. Choose δ' close to δ and p sufficiently large that $n/2 + R < \delta' + n/p < \delta$. It follows that $-\delta' < n/2 - n/p^* - R$, which implies as above that P has an infinite-dimensional kernel in $H^{m,p^*}_{-\delta'}(M;E)$. The fact that $C^{k-m,\alpha}_\delta(M;E) \subset H^{0,p}_{\delta'}(M;E)$ implies that $H^{0,p^*}_{-\delta'}(M;E) = (H^{0,p}_{\delta'}(M;E))^* \subset (C^{k-m,\alpha}_\delta(M;E))^*$. As above, each linear functional on $C^{k-m,\alpha}_\delta(M;E)$ defined by an element of $\mathrm{Ker}\, P \cap H^{m,p^*}_{-\delta'}(M;E)$ annihilates the range of P, so once again we conclude that $P\colon C^{k,\alpha}_\delta(M;E) \to C^{k-m,\alpha}_\delta(M;E)$ has infinite-dimensional cokernel.

Next we consider the borderline cases. The lower borderline Hölder case $\delta = n/2 - R$ was already treated above, when we showed that P has infinite-dimensional kernel in $C^{k,\alpha}_\delta(M;E)$ whenever $\delta \leq n/2 - R$. In all remaining cases, we will show that P does not have closed range.

We begin with the upper borderline Sobolev case, $H^{k,p}_{\delta_p}(M;E)$ with $\delta_p = n/2 - n/p + R$. Let us assume that $P: H^{k,p}_{\delta_p}(M;E) \to H^{k-m,p}_{\delta_p}(M;E)$ has closed range and derive a contradiction. If $\delta < \delta_p$ is chosen sufficiently close to δ_p that $|\delta - n/2 + n/p| < R$, the argument at the beginning of this proof showed that $\operatorname{Ker} P \cap H^{k,p}_\delta(M;E)$ is finite-dimensional. Since $H^{k,p}_{\delta_p}(M;E) \subset H^{k,p}_\delta(M;E)$, we see that $\operatorname{Ker} P \cap H^{k,p}_{\delta_p}(M;E)$ is finite-dimensional as well, so $P: H^{k,p}_{\delta_p}(M;E) \to H^{k-m,p}_{\delta_p}(M;E)$ is semi-Fredholm. It follows from Lemma 4.10 that there is a compact set $K \subset M$ and a constant C such that

$$\|u\|_{0,p,\delta_p} \leq C\|Pu\|_{0,p,\delta_p} \tag{6.29}$$

when $u \in H^{m,p}_{\delta_p}(M;E)$ is supported in $M \smallsetminus K$.

By definition of R, P has a characteristic exponent s_0 whose real part is equal to $n/2 - r + R$. Let w be any element of the space W defined in Lemma 6.7 corresponding to this characteristic exponent, with $\operatorname{supp} w \subset M \smallsetminus K$, $w \notin H^{0,p}_{\delta_p}(M;E)$, but $Pw \in C^{0,0}_{n/2+R+1}(M;E) \subset H^{0,p}_{\delta_p}(M;E)$. Let $\{\psi_\varepsilon\}$ be a family of cutoff functions as in Lemma 3.8, and define $w_\varepsilon = (1-\psi_\varepsilon)w$. Note that for any $\varepsilon > 0$, w_ε is in $C^{l,\beta}_{\mathrm{loc}}(M;E)$ and compactly supported, so it is in $H^{k,p}_{\delta_p}(M;E)$ for all $k \leq l$. Because $w \notin H^{0,p}_{\delta_p}(M;E)$ and $w_\varepsilon \to w$ uniformly on compact sets as $\varepsilon \to 0$, we have $\|w_\varepsilon\|_{0,p,\delta_p} \to \infty$. On the other hand,

$$Pw_\varepsilon = (1-\psi_\varepsilon)Pw - [P,\psi_\varepsilon]w \in H^{0,p}_{\delta_p}(M;E). \tag{6.30}$$

If we can show that $\|Pw_\varepsilon\|_{0,p,\delta_p}$ remains bounded as $\varepsilon \to 0$, we will have a contradiction to (6.29).

The fact that $Pw \in H^{0,p}_{\delta_p}(M;E)$ implies that $(1-\psi_\varepsilon)Pw \to Pw$ in the $H^{0,p}_{\delta_p}$ norm, so the first term in (6.30) is clearly bounded in $H^{0,p}_{\delta_p}$. For the second term, observe that the commutator $[\nabla,\psi_\varepsilon]w = w \otimes d\psi_\varepsilon$ is an operator of order zero with coefficients that are uniformly bounded in $C^{l,\beta}(M)$ and supported on the set where $\varepsilon/2 \leq \rho \leq \varepsilon$. It follows by induction that $[P,\psi_\varepsilon]$ is an operator of order $m-1$ with bounded coefficients supported in the same set, and therefore

$$\|[P,\psi_\varepsilon]w\|^p_{0,p,\delta_p} \leq C \sum_{0 \leq j \leq m-1} \int_{\varepsilon/2 \leq \rho \leq \varepsilon} |\nabla^j w|^p \, dV_g.$$

Since $|\nabla^j w|^p$ is integrable for $0 \leq j \leq m$, each integral above goes to zero as $\varepsilon \to 0$ by the dominated convergence theorem. This contradicts (6.29) and completes the proof that P does not have closed range in this case.

Next consider the upper borderline Hölder case, $\delta = n/2 + R$. The argument is almost the same as in the Sobolev case, except in this case we have to set $w_\varepsilon = \psi_\varepsilon w$ and show that $\|Pw_\varepsilon\|_{0,\alpha,n/2+R} \to 0$ while $\|w_\varepsilon\|_{0,\alpha,n/2+R}$ remains bounded below by a positive constant. The details are left to the reader.

The only case left is the lower borderline Sobolev case, $H^{k,p}_\delta(M;E)$ with $\delta = n/2-n/p-R$. Since $-\delta = n/2-n/p^*+R$, we showed above that $P: H^{m,p^*}_{-\delta}(M;E) \to H^{0,p^*}_{-\delta}(M;E)$ does not have closed range, and thus neither does $P: H^{0,p^*}_{-\delta}(M;E) \to H^{0,p^*}_{-\delta}(M;E)$ considered as an unbounded operator (since its domain is exactly

$H^{m,p^*}_{-\delta}(M;E)$). Since a closed, densely defined operator has closed range if and only if its adjoint does (cf. [**34**, Theorem IV.5.13]), this implies that $P\colon H^{0,p}_\delta(M;E) \to H^{0,p}_\delta(M;E)$ does not have closed range, and then it follows from the regularity results of Proposition 6.5 that $P\colon H^{k,p}_\delta(M;E) \to H^{k-m,p}_\delta(M;E)$ does not either.

Finally, to prove part (a), just observe that P being Fredholm on $L^2(M;E)$ implies that it is Fredholm from $H^{m,2}_0(M;E)$ to $H^{0,2}_0(M;E)$. By part (b), this in turn implies that $|0| < R$. □

CHAPTER 7

Laplace Operators

In this chapter we specialize to Laplace operators. Throughout this chapter, (M, g) will be a connected asymptotically hyperbolic $(n+1)$-manifold of class $C^{l,\beta}$ for some $l \geq 2$ and $0 \leq \beta < 1$.

Let E be a geometric tensor bundle of weight r over M. A *Laplace operator* is a second-order geometric operator $P \colon C^\infty(M; E) \to C^\infty(M; E)$ that can be written in the form $P = \nabla^*\nabla + \mathscr{K}$, where $\nabla^*\nabla$ is the covariant Laplacian and $\mathscr{K} \colon E \to E$ is a bundle endomorphism (i.e., a differential operator of order zero). Note that our definition of geometric operators guarantees that the coefficients of $\mathscr{K}$ in any local frame are contractions of tensor products of g, g^{-1}, dV_g, and the curvature tensor.

To get sharp Fredholm results for a specific Laplace operator, we need to compute its indicial radius. In general, this is just a straightforward computation in coordinates near the boundary. Here are some of the results.

LEMMA 7.1. *The covariant Laplacian $\nabla^*\nabla$ on trace-free symmetric r-tensors has indicial radius*

$$R = \sqrt{\frac{n^2}{4} + r}.$$

We will postpone the proof of this lemma until after Proposition 7.3 below. For the record, we also note the following, which is proved in [**39, 40**].

LEMMA 7.2 (Mazzeo). *The Laplace-Beltrami operator $\Delta = dd^* + d^*d$ on q-forms has the following indicial radius:*

$$R = \begin{cases} \dfrac{n}{2} - q, & 0 \leq q \leq \dfrac{n}{2}, \\ \dfrac{1}{2}, & q = \dfrac{n+1}{2}, \\ q - \dfrac{n+2}{2}, & \dfrac{n+2}{2} \leq q \leq n+1. \end{cases}$$

The following lemma greatly simplifies the computation of the indicial radius of a Laplace operator.

PROPOSITION 7.3. *Let $P = \nabla^*\nabla + \mathscr{K}$ be a Laplace operator acting on a geometric tensor bundle of weight r. For any $s \in \mathbb{C}$,*

$$I_s(P) = I_0(P) + s(n - s - 2r).$$

PROOF. Since $I_s(\mathscr{K}) = \mathscr{K}|_{\partial M}$ is independent of s, we need only consider the case $P = \nabla^*\nabla$. As in the proof of Proposition 2.7 of [**29**], we compute (in background coordinates)

$$\begin{aligned}
\rho_{;ij} &= \partial_i \partial_j \rho - \Gamma^k_{ij} \partial_k \rho \\
&= \rho^{-1}(2\partial_i \rho \partial_j \rho - \overline{g}_{ij}) + o(\rho^{-1}); \\
\Delta(\rho^s) &= -s(s-1)\rho^{s-2} g^{jk} \rho_{;j} \rho_{;k} - s\rho^{s-1} \rho_{;k}{}^k \\
&= s(n-s)\rho^s + o(\rho^s); \\
|\nabla \rho|^2_g &= |d\rho|^2_g \\
&= \rho^2 |d\rho|^2_{\overline{g}} = \rho^2 + o(\rho^2).
\end{aligned} \qquad (7.1)$$

Therefore,

$$\begin{aligned}
\rho^{-s} \nabla^* \nabla(\rho^s \overline{u}) &= -\rho^{-s} \operatorname{Tr}_g \nabla^2(\rho^s \overline{u}) \\
&= -\rho^{-s} \operatorname{Tr}_g \nabla(s\rho^{s-1} \overline{u} \otimes \nabla \rho + \rho^s \nabla \overline{u}) \\
&= -\rho^{-s} \operatorname{Tr}_g \left(s(s-1)\rho^{s-2} \overline{u} \otimes \nabla \rho \otimes \nabla \rho + 2s\rho^{s-1} \nabla \overline{u} \otimes \nabla \rho \right. \\
&\qquad \left. + s\rho^{s-1} \overline{u} \otimes \nabla^2 \rho + \rho^s \nabla^2 \overline{u} \right) \\
&= -s(s-1)\rho^{-2} |\nabla \rho|^2_g \overline{u} - 2s\rho^{-1} \nabla_{\operatorname{grad}\rho} \overline{u} \\
&\qquad + s\rho^{-1} \Delta \rho \overline{u} + \nabla^* \nabla \overline{u} \\
&= -s(s-1)\overline{u} - 2s\rho^{-1} \nabla_{\operatorname{grad}\rho} \overline{u} + s(n-1)\overline{u} + \nabla^* \nabla \overline{u} + o(1).
\end{aligned} \qquad (7.2)$$

To compute the second term above, assume $\overline{u}$ is a tensor of type $\binom{q}{p}$ with $q-p=r$, and let D be the difference tensor $D = \nabla - \overline{\nabla}$. The components of $\nabla \overline{u}$ are

$$\begin{aligned}
\overline{u}^{i_1...i_p}_{j_1...j_q;k} &= \partial_k \overline{u}^{i_1...i_p}_{j_1...j_q} + \sum_{s=1}^p \Gamma^{i_s}_{mk} \overline{u}^{i_1...m...i_p}_{j_1...j_q} - \sum_{s=1}^q \Gamma^m_{j_s k} \overline{u}^{i_1...i_p}_{j_1...m...j_q} \\
&= O(1) + \sum_{s=1}^p D^{i_s}_{mk} \overline{u}^{i_1...m...i_p}_{j_1...j_q} - \sum_{s=1}^q D^m_{j_s k} \overline{u}^{i_1...i_p}_{j_1...m...j_q}.
\end{aligned}$$

Let us introduce the shorthand notations $\rho_i = \partial_i \rho$ and $\overline{\rho}^i = \overline{g}^{ij} \partial_j \rho$. Using formula (3.10) for the components of D, together with the fact that $\rho_i \overline{\rho}^i = 1 + O(\rho)$, we obtain

$$\rho^{-1} g^{kl} \rho_l D^i_{jk} = -\overline{\rho}^k (\delta^i_j \rho_k + \delta^i_k \rho_j - \overline{g}_{jk} \overline{\rho}^i) = -\delta^i_j + O(\rho).$$

Therefore,

$$\begin{aligned}
\rho^{-1}(\nabla_{\operatorname{grad}\rho} \overline{u})^{i_1...i_p}_{j_1...j_q} &= \rho^{-1} g^{kl} \rho_l \overline{u}^{i_1...i_p}_{j_1...j_q;k} \\
&= -\sum_{s=1}^p \delta^{i_s}_m \overline{u}^{i_1...m...i_p}_{j_1...j_q} + \sum_{s=1}^q \delta^m_{j_s} \overline{u}^{i_1...i_p}_{j_1...m...j_q} + O(\rho) \\
&= r \overline{u}^{i_1...i_p}_{j_1...j_q} + O(\rho).
\end{aligned}$$

Inserting this back into (7.2), we obtain

$$\rho^{-s} \nabla^* \nabla(\rho^s \overline{u}) = s(n-s-2r)\overline{u} + \nabla^* \nabla \overline{u} + o(1),$$

which implies the result. $\square$

PROOF OF LEMMA 7.1. By Prop. 7.3, to determine $I_s(\nabla^* \nabla)$ it suffices to compute $I_0(\nabla^* \nabla)$. The cases $r=0$ and $r=2$ follow from Corollary 2.8 and Lemma 2.9 of [29]. For the general case, suppose $r \geq 1$. Assuming $\overline{u}$ is trace-free, symmetric,

7. LAPLACE OPERATORS

and smooth up to the boundary, and using the notation of the preceding proof, we compute

$$(\nabla^*\nabla \overline{u})_{i_1\ldots i_r} = -\overline{u}_{i_1\ldots i_r;l}{}^l$$
$$= -g^{lm}\left(\partial_m \overline{u}_{i_1\ldots i_r;l} - D_{lm}^j \overline{u}_{i_1\ldots i_r;j} - \sum_{s=1}^r D_{i_s m}^j \overline{u}_{i_1\ldots j\ldots i_r;l}\right) + O(\rho).$$

As in the preceding proof, we have

$$\overline{u}_{i_1\ldots i_r;l} = -\sum_{t=1}^r D_{i_t l}^k \overline{u}_{i_1\ldots k\ldots i_r} + O(1),$$

and therefore,

$$(\nabla^*\nabla \overline{u})_{i_1\ldots i_r} = \sum_{t=1}^r g^{lm}(\partial_m D_{i_t l}^k)\overline{u}_{i_1\ldots k\ldots i_r} - \sum_{t=1}^r g^{lm}D_{lm}^j D_{i_t j}^k \overline{u}_{i_1\ldots k\ldots i_r}$$
$$-\sum_{\substack{1\leq s,t\leq r\\ s\neq t}} g^{lm}D_{i_s m}^j D_{i_t l}^k \overline{u}_{i_1\ldots j\ldots k\ldots i_r} - \sum_{s=1}^r g^{lm}D_{i_s m}^j D_{jl}^k \overline{u}_{i_1\ldots k\ldots i_r}$$
$$+ O(\rho).$$

Using (3.10) again, we compute

$$g^{lm}\partial_m D_{i_t l}^k = g^{lm}\rho^{-2}\rho_m\left(\delta_{i_t}^k \rho_l + \delta_l^k \rho_{i_t} - \overline{g}_{i_t l}\overline{\rho}^k\right) + O(\rho) = \delta_{i_t}^k + O(\rho);$$
$$g^{lm}D_{lm}^j D_{i_t j}^k = -(n-1)\delta_{i_t}^k + O(\rho);$$
$$g^{lm}D_{i_s m}^j D_{i_t l}^k = \delta_{i_s}^j \delta_{i_t}^k + \overline{g}^{jk}\rho_{i_s}\rho_{i_t} - \delta_{i_s}^k \rho_{i_t}\overline{\rho}^j - \delta_{i_t}^j \rho_{i_s}\overline{\rho}^k + \overline{g}_{i_s i_t}\overline{\rho}^j\overline{\rho}^k + O(\rho);$$
$$g^{lm}D_{i_s m}^j D_{jl}^k = -(n-1)\rho_{i_s}\overline{\rho}^k + O(\rho).$$

Inserting these above and using the fact that $\overline{u}$ is symmetric and trace-free, we obtain

$$(\nabla^*\nabla \overline{u})_{i_1\ldots i_r} = r\overline{u}_{i_1\ldots i_r} + r(n-1)\overline{u}_{i_1\ldots i_r} - r(r-1)\overline{u}_{i_1\ldots i_r} - 0$$
$$+ (r-1)\sum_{t=1}^r \rho_{i_t}\overline{\rho}^j \overline{u}_{i_1\ldots j\ldots i_r} + (r-1)\sum_{s=1}^r \rho_{i_s}\overline{\rho}^k \overline{u}_{i_1\ldots k\ldots i_r}$$
$$-\sum_{\substack{1\leq s,t\leq r\\ s\neq t}}\overline{g}_{i_s i_t}\overline{\rho}^j\overline{\rho}^k \overline{u}_{i_1\ldots j\ldots k\ldots i_r} + (n-1)\sum_{s=1}^r \rho_{i_s}\overline{\rho}^k \overline{u}_{i_1\ldots k\ldots i_r} + O(\rho)$$
$$= r(n+1-r)\overline{u}_{i_1\ldots i_r} + (2r+n-3)\sum_{t=1}^r \rho_{i_t}\overline{\rho}^j \overline{u}_{i_1\ldots j\ldots i_r}$$
$$-\sum_{\substack{1\leq s,t\leq r\\ s\neq t}}\overline{g}_{i_s i_t}\overline{\rho}^j\overline{\rho}^k \overline{u}_{i_1\ldots j\ldots k\ldots i_r} + O(\rho).$$

Suppose s is any indicial root of $\nabla^*\nabla$ and $\overline{u}$ is a corresponding unit eigentensor. Using the formula above for $\nabla^*\nabla \overline{u}$ together with Proposition 7.3, and observing that $\overline{u}$ is trace-free, we compute

$$0 = \langle \overline{u}, I_s(\nabla^*\nabla)\overline{u}\rangle_{\overline{g}}$$
$$= r(n+1-r)|\overline{u}|_{\overline{g}}^2 + (2r+n-3)\left|\operatorname{grad}_{\overline{g}}\rho \lrcorner \overline{u}\right|_{\overline{g}}^2 + s(n-s-2r)|\overline{u}|_{\overline{g}}^2$$
$$\geq r(n+1-r) + s(n-s-2r),$$

which implies that each indicial root satisfies

$$\left|s - \left(\frac{n}{2} - r\right)\right|^2 \geq \frac{n^2}{4} + r.$$

It follows that the indicial radius of $\nabla^*\nabla$ is at least $\sqrt{n^2/4 + r}$.

On the other hand, if $\overline{u}$ is chosen so that $\operatorname{grad}_{\overline{g}}\rho \lrcorner \overline{u} = 0$ along ∂M (i.e., $\overline{u}$ is purely tangential), we find that $I_s(\nabla^*\nabla)\overline{u} = \bigl(r(n+1-r) + s(n-s-2r)\bigr)\overline{u}$. Solving $I_s(\nabla^*\nabla)\overline{u} = 0$ for s, we find that two of the indicial roots of $\nabla^*\nabla$ are

$$s = \frac{n}{2} - r \pm \sqrt{\frac{n^2}{4} + r}.$$

This proves that the indicial radius is exactly $\sqrt{n^2/4 + r}$ as claimed. □

COROLLARY 7.4. *Let $P = \nabla^*\nabla + \mathscr{K}$ be a Laplace operator acting on a geometric tensor bundle of weight r, and suppose P has indicial radius $R > 0$. For $c \in \mathbb{R}$, the indicial radius R' of $P + c$ is positive if and only if $c + R^2 > 0$, in which case*

$$R' = \sqrt{c + R^2}.$$

PROOF. Comparing the formulas given by Proposition 7.3 for $I_s(P)$ and $I_{n/2-r}(P)$, we find that

$$I_s(P) = I_{n/2-r}(P) - (s - n/2 + r)^2.$$

Observe that $I_{n/2-r}(P)$ is self-adjoint by Proposition 4.4, so it has real eigenvalues. Now s is a characteristic root of P precisely when s is a solution to the quadratic equation $(s - n/2 + r)^2 = \mu$ for some eigenvalue μ of $I_{n/2-r}(P)$. If some eigenvalue were nonpositive, this equation would have a root with real part equal to $n/2 - r$, which would imply $R = 0$. Therefore, the assumption $R > 0$ means that all of the eigenvalues $\{\mu_i\}$ of $I_{n/2-r}(P)$ are strictly positive, and therefore the characteristic roots of P are $s = n/2 - r \pm \sqrt{\mu_i}$, with $R^2 = \min\{\mu_i\}$.

Since $I_s(P + c) = I_s(P) + c$, the characteristic roots of $P + c$ are $s = n/2 - r \pm \sqrt{c + \mu_i}$, and the one with smallest real part greater than $n/2 - r$ is $s = n/2 - r \pm \sqrt{c + R^2}$. Thus the indicial radius of $P + c$ is $\sqrt{c + R^2}$ as claimed. □

LEMMA 7.5. *Let Δ_L be the Lichnerowicz Laplacian, and let c be a real constant. If $n^2/4 - 2n + c > 0$, then the indicial radius of $\Delta_L + c$ acting on symmetric 2-tensors is*

$$R = \sqrt{\frac{n^2}{4} - 2n + c}. \tag{7.3}$$

PROOF. Observe first that Δ_L preserves the splitting of symmetric 2-tensors into trace and trace-free parts:

$$\Sigma^2 M = \mathbb{R}g \oplus \Sigma_0^2 M,$$

where $\Sigma^2 M$ is the bundle of symmetric covariant 2-tensors, $\Sigma_0^2 M$ is the subbundle of tensors that are trace-free with respect to g, and $\mathbb{R}g \subset \Sigma^2 M$ is the real line

bundle of multiples of g. On $\mathbb{R}g$, $\mathring{Rc}(ug) = \mathring{Rm}(ug)$, so Δ_L acts as the ordinary Laplacian:
$$\Delta_L(ug) = (\nabla^*\nabla u)g.$$
It follows from Lemma 7.1 (or Lemma 7.2) that the indicial radius of $\nabla^*\nabla$ on functions is $R = n/2$, so the indicial radius of $\nabla^*\nabla + c$ is $\sqrt{n^2/4 + c}$, which is greater than (7.3). Therefore, it suffices to show that $\Delta_L + c$ acting on trace-free symmetric 2-tensors has indicial radius given by (7.3).

The asymptotic formula (4.3) for the Riemann curvature tensor implies that the action of $\mathring{Rm}$ and $\mathring{Rc}$ on $\Sigma_0^2 M$ near the boundary is given by
$$\begin{aligned}\mathring{Rc}(u) &= -nu + O(\rho|u|); \\ \mathring{Rm}(u) &= u + O(\rho|u|).\end{aligned} \quad (7.4)$$

Thus the indicial radius of $\Delta_L + c$ on $\Sigma_0^2 M$ is the same as that of $\nabla^*\nabla - 2n - 2 + c$, which is $\sqrt{n^2/4 - 2n + c}$ by Lemma 7.1 and Corollary 7.4. □

The main thing that needs to be checked in order to apply Theorem C is the L^2 estimate (1.4). For some operators, an appropriate asymptotic estimate follows from an obvious integration by parts, such as $\nabla^*\nabla + c$ when c is a positive constant:
$$(u, \nabla^*\nabla u + cu) = \|\nabla u\|^2 + c\|u\|^2 \geq c\|u\|^2,$$
from which $\|u\| \leq c^{-1}\|(\nabla^*\nabla + c)u\|$ follows by the Cauchy-Schwartz inequality. However, when the zero-order term is not strictly positive, we need to work a bit harder.

As a warmup for the general L^2 estimates we will prove below, consider first the ordinary Laplacian $\Delta = d^*d$ on functions. The use of positive eigenfunctions, and more generally positive functions satisfying differential inequalities, is a common tool for estimating the lower bound of the spectrum of elliptic operators; see for example [20, 38, 43] and especially [56], where such functions play a central role. The following lemma was proved originally by Cheng and Yau [20, p. 345].

LEMMA 7.6 (Cheng-Yau). *Let M be any Riemannian manifold. If there exists a positive, locally C^2 function φ on M such that $\Delta\varphi/\varphi \geq \lambda$, then*
$$(u, \Delta u) \geq \lambda\|u\|^2 \quad (7.5)$$
for all smooth compactly supported functions u.

The proof of this lemma in [20] uses the maximum principle. Here is a simple proof based on integration by parts, which serves to motivate the somewhat more delicate estimates below.

PROOF. Let $u \in C_c^\infty(M)$. The divergence theorem gives
$$\begin{aligned}0 &= \int_M d^*(u^2\varphi^{-1}d\varphi)\,dV_g \\ &= \int_M \left(-2u\varphi^{-1}\langle du, d\varphi\rangle_g + u^2\varphi^{-2}|d\varphi|^2 + u^2\varphi^{-1}\Delta\varphi\right)dV_g.\end{aligned}$$

Thus

$$0 \leq \int_M \left|\varphi d(\varphi^{-1}u)\right|^2 dV_g$$
$$= \int_M \left|du - u\varphi^{-1}d\varphi\right|^2 dV_g$$
$$= \int_M \left(|du|^2 - 2u\varphi^{-1}\langle du, d\varphi\rangle_g + u^2\varphi^{-2}|d\varphi|^2\right) dV_g$$
$$= \int_M \left(u\Delta u - u^2\varphi^{-1}\Delta\varphi\right) dV_g \leq \int_M \left(u\Delta u - \lambda u^2\right) dV_g,$$

which is equivalent to (7.5). □

Lemma 7.6 is a global result; but for proving asymptotic estimates, we need only find a function φ that satisfies $\Delta\varphi/\varphi \geq \lambda$ on the complement of a compact set. In particular, on an asymptotically hyperbolic $(n+1)$-manifold, asymptotic computations show that $\Delta(\rho^{n/2})/\rho^{n/2}$ can be made arbitrarily close to $n^2/4$ near ∂M. From this it will follow, for example, that $\Delta - \lambda$ on functions satisfies the hypotheses of Theorem C for any constant $\lambda < n^2/4$.

It is interesting to note that this simple estimate immediately yields a (rather crude) estimate for the covariant Laplacian on tensor fields.

LEMMA 7.7. *If (7.5) holds for smooth functions compactly supported in some open set $U \subset M$, then for any smooth tensor field w compactly supported in U, we have*

$$(w, \nabla^*\nabla w) \geq \lambda\|w\|^2$$

with the same constant λ.

PROOF. Inequality (7.5) implies

$$\lambda\|u\|^2 \leq \|\nabla u\|^2,$$

which extends continuously to compactly supported functions in $H^{1,2}(U)$. If w is a smooth, compactly supported tensor field, the function $|w|$ is Lipschitz, hence in $H^{1,2}(U)$. Kato's inequality says that $|\nabla|w|| \leq |\nabla w|$ almost everywhere (see [**10**, Prop. 3.49], where this is proved for scalar functions; the proof extends easily to tensors). Therefore

$$\lambda\|w\|^2 \leq \|\nabla|w|\|^2 \leq \|\nabla w\|^2 = (w, \nabla^*\nabla w).$$

□

A version of this argument (using the maximum principle instead of integration by parts) was used implicitly in our proof that the linearized Einstein operator is invertible on weighted Hölder spaces over hyperbolic space [**29**]. Unfortunately, as we noted there, this estimate was not sharp, and led to less-than-optimal Fredholm results for tensors. For the application to Einstein metrics in this monograph, we no longer need a sharp asymptotic estimate, because of the sharp Fredholm theorems of the preceding chapter. However, with other applications in mind, it is useful to see how far the asymptotic estimates can be pushed.

The key to finding improved asymptotic estimates on tensors turns out to be to consider r-tensor fields as $(r-1)$-tensor-valued 1-forms. Therefore we must make a short digression to discuss the properties of tensor-valued differential forms.

7. LAPLACE OPERATORS

Let E be any geometric tensor bundle over M, and let $\Lambda^q E := E \otimes \Lambda^q M$ denote the bundle of E-valued q-forms on M. Let $D: C^\infty(M; \Lambda^q E) \to C^\infty(M; \Lambda^{q+1} E)$ denote the exterior covariant differential on E-valued forms, defined by

$$D(\sigma \otimes \alpha) = \nabla\sigma \wedge \alpha + \sigma \otimes d\alpha,$$

for $\alpha \in C^\infty(M; \Lambda^q M)$ and $\sigma \in C^\infty(M; E)$. (The wedge product above is computed by wedging the 1-form component of $\nabla\sigma$ with α to yield a section of $\Lambda^{q+1} E$.) We will study the covariant Laplace-Beltrami operator on E-valued forms, defined by $\Delta = DD^* + D^*D$, where D^* is the formal adjoint of D.

For a scalar 1-form α, we let $\alpha \vee : \Lambda^{q+1} E \to \Lambda^q E$ denote the (pointwise) adjoint of the operator $\alpha \wedge : \Lambda^q E \to \Lambda^{q+1} E$ with respect to g, so that $\langle \alpha \wedge \omega, \eta \rangle_g = \langle \omega, \alpha \vee \eta \rangle_g$. In particular, if β is also a scalar 1-form, then $\alpha \vee \beta = \langle \alpha, \beta \rangle_g$.

For any function $u \in C^2(M)$, let $H(u)$ denote the covariant Hessian of u acting as bundle endomorphism $H(u): \Lambda^1 M \to \Lambda^1 M$, and extended to $\Lambda^q M$ as a derivation. In terms of any orthonormal basis,

$$H(u)\omega = u_{;ij} e^i \wedge (e^j \vee \omega) \tag{7.6}$$

(where $u_{;ij}$ are the components of $\nabla^2 u$), since both sides are derivations that agree on $\Lambda^1 M$. We extend this endomorphism to $\Lambda^q E$ by letting it act on the differential form component alone.

LEMMA 7.8. *Suppose α and β are scalar 1-forms, ω is an E-valued q-form, and u is a function. For any local orthonormal frame $\{e_j\}$ for TM and dual coframe $\{e^j\}$, we have the following facts:*
  (a) $D\omega = e^j \wedge \nabla_{e_j}\omega$.
  (b) $D^*\omega = -e^j \vee \nabla_{e_j}\omega$.
  (c) $D(u\omega) = uD\omega + du \wedge \omega$.
  (d) $D^*(u\omega) = uD^*\omega - du \vee \omega$.
  (e) $\alpha \wedge (\beta \vee \omega) + \beta \vee (\alpha \wedge \omega) = \langle \alpha, \beta \rangle_g \omega$.
  (f) $D(du \vee \omega) = -du \vee D\omega + H(u)\omega + \nabla_{\mathrm{grad}\, u}\omega$.
  (g) $D^*(du \wedge \omega) = -du \wedge D^*\omega + H(u)\omega - \nabla_{\mathrm{grad}\, u}\omega + (\Delta u)\omega$.

PROOF. Parts (a) through (e) are standard, and can be found, for example, in [**60**, Ch. 2 and Section 6.1]. For (f), choose a point $p \in M$ and a frame $\{e_j\}$ such that $\nabla e_j = \nabla e^j = 0$ at p. Then, computing at p and using (a), (e), and (7.6), we have

$$\begin{aligned}
D(du \vee \omega) &= e^j \wedge \nabla_j(u_{;k} e^k \vee \omega) \\
&= e^j \wedge (u_{;jk} e^k \vee \omega) + u_{;k} e^j \wedge (e^k \vee \nabla_j \omega) \\
&= H(u)\omega + u_{;k} g^{jk} \nabla_j \omega - u_{;k} e^k \vee (e^j \wedge \nabla_j \omega) \\
&= H(u)\omega + \nabla_{\mathrm{grad}\, u}\omega - du \vee D\omega.
\end{aligned}$$

The computation for (g) is similar. □

The integral formula in the next lemma, a tensor analogue of (7.5), is the key to proving sharp asymptotic L^2 estimates. It unifies and generalizes Bochner-type formulas that have been used in various settings, such as the weighted Bochner formula introduced by Witten [**58**] to prove the Morse inequalities, and a closely related formula for scalar differential forms used by Donnelly and Xavier [**25**] to analyze the spectrum of the scalar Laplace-Beltrami operator on negatively-curved manifolds, and by Lars Andersson [**7**] for the same operator on asymptotically hyperbolic manifolds. Later Andersson and Chruściel [**8**] used the formula presented here (based on an early draft of the present monograph) to obtain Fredholm results for the "vector Laplacian" L^*L that arises in the constraint equations of general relativity (see the Introduction).

LEMMA 7.9. *For any smooth, compactly supported section ω of $\Lambda^q E$, and any positive C^2 function φ on M, the following integral formula holds:*

$$\begin{aligned}(\omega, \Delta\omega) &= \int_M \varphi^{-1}\Delta\varphi|\omega|_g^2 + 2\left\langle\omega, H(\log\varphi)\omega\right\rangle_g \\ &\quad + |\varphi D(\varphi^{-1}\omega)|_g^2 + |\varphi^{-1}D^*(\varphi\omega)|_g^2 \, dV_g \\ &\geq \int_M \left\langle\omega, (\varphi^{-1}\Delta\varphi + 2H(\log\varphi))\omega\right\rangle_g \, dV_g.\end{aligned} \quad (7.7)$$

PROOF. Let $u = \log\varphi$. Using Lemma 7.8, we compute

$$\begin{aligned}&\int_M |e^u D(e^{-u}\omega)|_g^2 + |e^{-u}D^*(e^u\omega)|_g^2 \, dV_g \\ &= \int_M |D\omega - du\wedge\omega|_g^2 + |D^*\omega - du\vee\omega|_g^2 \, dV_g \\ &= \int_M |D\omega|_g^2 - 2\left\langle D\omega, du\wedge\omega\right\rangle_g + |du\wedge\omega|_g^2 \\ &\quad + |D^*\omega|_g^2 - 2\left\langle D^*\omega, du\vee\omega\right\rangle_g + |du\vee\omega|_g^2 \, dV_g \\ &= \int_M \left\langle\omega, \Delta\omega\right\rangle_g - 2\left\langle du\vee D\omega, \omega\right\rangle_g + \left\langle du\vee(du\wedge\omega), \omega\right\rangle_g \\ &\quad - 2\left\langle\omega, D(du\vee\omega)\right\rangle_g + \left\langle du\wedge(du\vee\omega), \omega\right\rangle_g \, dV_g \\ &= \int_M \left\langle\omega, \Delta\omega\right\rangle_g + |du|_g^2|\omega|_g^2 - 2\left\langle\omega, H(u)\omega\right\rangle_g - 2\left\langle\omega, \nabla_{\operatorname{grad} u}\omega\right\rangle_g \, dV_g \\ &= \int_M \left\langle\omega, \Delta\omega\right\rangle_g + |du|_g^2|\omega|_g^2 - 2\left\langle\omega, H(u)\omega\right\rangle_g - (\Delta u)|\omega|_g^2 \, dV_g \\ &= \int_M \left\langle\omega, \Delta\omega\right\rangle_g - e^{-u}\Delta(e^u)|\omega|_g^2 - 2\left\langle\omega, H(u)\omega\right\rangle_g \, dV_g.\end{aligned}$$

□

To make use of this formula, we will use a power of the defining function ρ as our weight function φ. It is convenient to introduce the following notation: If $P\colon C^\infty(M;E) \to C^\infty(M;E)$ is a differential operator on M and λ is a real number, we write

$$(u, Pu) \gtrsim \lambda \|u\|^2 \quad (7.8)$$

to mean that for every $\varepsilon > 0$, there exists a compact set K_ε such that

$$(u, Pu) \geq (\lambda - \varepsilon)\|u\|^2$$

whenever u is smooth and compactly supported in $M \smallsetminus K_\varepsilon$.

LEMMA 7.10. *Let M be an asymptotically hyperbolic $(n+1)$-manifold of class $C^{l,\beta}$, with $l \geq 2$ and $0 \leq \beta < 1$, and suppose either $0 \leq q < n/2$ or $(n+2)/2 <$*

$q \leq n$. The covariant Laplace-Beltrami operator satisfies the following asymptotic estimate on E-valued q-forms:

$$(\omega, \Delta\omega) \gtrsim \lambda \|\omega\|^2,$$

where $\lambda = (n-2q)^2/4$ if $0 \leq q < n/2$ and $\lambda = (n+2-2q)^2/4$ if $(n+2)/2 < q \leq n$.

PROOF. We will use (7.7) for forms ω supported near the boundary. Let ρ be a smooth defining function. Using (7.1), we compute

$$(\log \rho)_{;ij} = \rho^{-2} \rho_{;i} \rho_{;j} - g_{ij} + O(\rho).$$

The tensor g acts as the identity on 1-forms, and since we extend it to act as a derivation, it acts on E-valued q-forms as q times the identity. Using (7.6), therefore, the action of $H(\log \rho)$ on an E-valued q-form ω can be written

$$\langle \omega, H(\log \rho)\omega \rangle_g = \langle \omega, (\log \rho)_{;ij} e^i \wedge (e^j \vee \omega) \rangle_g$$

$$= \left\langle \omega, \frac{d\rho}{\rho} \wedge \left(\frac{d\rho}{\rho} \vee \omega \right) \right\rangle_g - q|\omega|_g^2 + O(\rho|\omega|_g^2)$$

$$= \left| \frac{d\rho}{\rho} \vee \omega \right|_g^2 - q|\omega|_g^2 + O(\rho|\omega|_g^2).$$

Thus with $\varphi = \rho^s$, where s is a constant to be determined later, the integrand on the right-hand side of (7.7) can be estimated as follows:

$$\langle \omega, (\rho^{-s} \Delta(\rho^s) + 2sH(\log \rho))\omega \rangle$$
$$\geq s(n-s-2q)|\omega|_g^2 + 2s \left| \frac{d\rho}{\rho} \vee \omega \right|_g^2 - O(\rho|\omega|_g^2). \quad (7.9)$$

Given $\delta > 0$, we can choose ε small enough that the absolute value of the $O(\rho|\omega|_g^2)$ factor above is bounded by $\delta|\omega|_g^2$ on A_ε. If $s \geq 0$, we then have for ω compactly supported in A_ε

$$(\omega, \Delta\omega) \geq (s(n-s-2q) - \delta)\|\omega\|_g^2.$$

This estimate is optimal when $s = (n-2q)/2$, so as long as $(n-2q)/2 > 0$, which is to say $q < n/2$, we obtain the conclusion of the lemma with $\lambda = (n-2q)^2/4$.

If on the other hand $s \leq 0$, we use the fact that $|d\rho/\rho|_g$ approaches 1 uniformly at ∂M. We choose ε small enough that the $O(\rho|\omega|_g^2)$ factor in (7.9) is bounded by $(\delta/2)|\omega|_g^2$ and $2s|d\rho/\rho|_g^2 \geq 2s - \delta/2$ on A_ε, and conclude that

$$(\omega, \Delta\omega) \geq (s(n-s-2q) - \delta/2)\|\omega\|^2 + (2s - \delta/2)\|\omega\|^2$$
$$= (s(n-s-2q+2) - \delta)\|\omega\|^2.$$

This in turn is optimal when $s = (n+2-2q)/2$. Thus as long as $q > (n+2)/2$, so that $s < 0$, we obtain the conclusion of the lemma with $\lambda = (n+2-2q)^2/4$. $\square$

When applied to scalar-valued forms, this result can be used to obtain an elementary proof of the sharp L^2 Fredholm theorems and spectral bounds for the Laplace-Beltrami operator originally obtained by Mazzeo [39, 40]. (See [7], where this is carried out in detail.) Our main interest in this monograph, however, is in the covariant Laplacian acting on symmetric tensors. In this case, we consider the bundle $\Sigma^r M$ of symmetric r-tensors as a subbundle of the bundle $\Lambda^1 T^{r-1} M$ of $(r-1)$-tensor-valued 1-forms.

The following lemma gives a Weitzenböck formula relating the covariant Laplacian on such tensors to Δ. For this purpose, we define zero-order operators $\widetilde{Rc}(u)$ and $\widetilde{Rm}(u)$ acting on r-tensors by

$$\widetilde{Rc}(u)_{i_1\ldots i_{r-1}j} = R_j{}^k u_{i_1\ldots i_{r-1}k};$$

$$\widetilde{Rm}(u)_{i_1\ldots i_{r-1}j} = \sum_{p=1}^{r-1} R_{i_p}{}^l{}_j{}^k u_{i_1\ldots l\ldots i_{r-1}k}.$$

Note that on symmetric 2-tensors, $\widetilde{Rm}$ agrees with the operator $\mathring{Rm}$ defined by (1.1), but $\widetilde{Rc}$ is not the same as $\mathring{Rc}$ in general.

LEMMA 7.11. *For a section u of $\Lambda^1 T^{r-1} M$,*

$$\Delta u = \nabla^*\nabla u + \widetilde{Rc}(u) - \widetilde{Rm}(u). \tag{7.10}$$

PROOF. This is easiest to see in components, noting that the *last* index of u is considered to be the 1-form index.

$$(D^*Du)_{i_1\ldots i_{r-1}j} = -u_{i_1\ldots i_{r-1}j;k}{}^k + u_{i_1\ldots i_{r-1}k;j}{}^k;$$

$$(DD^*u)_{i_1\ldots i_{r-1}j} = -u_{i_1\ldots i_{r-1}k;}{}^k{}_j.$$

Applying the Ricci identity to the commutator

$$u_{i_1\ldots i_{r-1}k;j}{}^k - u_{i_1\ldots i_{r-1}k;}{}^k{}_j$$

yields the result. □

Specializing Lemma 7.10 to $\Sigma_0^r M$, we obtain the following sharp asymptotic estimate.

LEMMA 7.12. *The following asymptotic estimate holds for any smooth, compactly supported, trace-free symmetric r-tensor u:*

$$(u, \nabla^*\nabla u) \gtrsim \left(\frac{n^2}{4} + r\right) \|u\|^2.$$

PROOF. Using (4.3), we compute that the curvature operators $\widetilde{Rc}$ and $\widetilde{Rm}$ have the following asymptotic behavior on trace-free symmetric r-tensors near ∂M:

$$\widetilde{Rc}(u) = -nu + O(\rho|u|_g);$$
$$\widetilde{Rm}(u) = (r-1)u + O(\rho|u|_g).$$

To prove the lemma, just use Lemma 7.10 with $q = 1$ together with Lemma 7.11 to obtain

$$(u, \nabla^*\nabla u) = (u, \Delta u) - (u, \widetilde{Rc}(u)) + (u, \widetilde{Rm}(u))$$
$$\gtrsim \frac{(n-2)^2}{4}\|u\|^2 + n\|u\|^2 + (r-1)\|u\|^2$$
$$= \left(\frac{n^2}{4} + r\right)\|u\|^2.$$

□

REMARK. The same method yields similar estimates for Laplace operators acting on any other tensor bundle, but in cases other than fully symmetric or fully antisymmetric tensors, the estimates so obtained appear not to be sharp. It would be interesting to know if a modified version of (7.7) could be used to sharpen the estimate in those cases.

LEMMA 7.13. *The following asymptotic estimate holds for any smooth, compactly supported, trace-free symmetric 2-tensor u:*

$$(u, \Delta_L u) \gtrsim \left(\frac{n^2}{4} - 2n\right) \|u\|^2.$$

PROOF. This follows immediately from the preceding lemma together with the asymptotic formulas (7.4) for $\mathring{Rm}$ and $\mathring{Rc}$. □

PROOF OF PROPOSITIONS D, E, F, AND G. The operators $\Delta_L + c$ and $\nabla^*\nabla + c$ are Fredholm on L^2 for the claimed values of c by virtue of Lemmas 7.13 and 7.12. On the other hand, the indicial radius computations at the beginning of this chapter show that these are precisely the values of c for which these operators have positive indicial radius, so these are the only values of c for which the operators are Fredholm. The result for the Hodge Laplacian follows similarly from Lemma 7.10 in the case of scalar forms. The claims about the essential spectrum follow immediately from the Fredholm results: Since each of these operators is self-adjoint on L^2, its spectrum is contained in $\mathbb{R}$, and a real number λ is in the essential spectrum if and only if $P - \lambda: L^2(M; E) \to L^2(M; E)$ is not Fredholm.

For the vector Laplacian L^*L, L. Andersson proved in [**7**, Lemma 3.15] that the following asymptotic estimate holds:

$$(V, L^*LV) \gtrsim \frac{n^2}{8} \|V\|^2.$$

It follows from Proposition B that L^*L is Fredholm on L^2, and then the rest of the claims follow from Theorem C. □

We conclude this chapter by observing that the sharp a priori L^2 estimates developed here for Laplace operators actually lead directly to sharp L^2 Fredholm theorems for such operators, without any need for the parametrix construction of the preceding chapter. This follows from the simple device of replacing u by $\rho^{-\delta}u$ to convert unweighted L^2 estimates to weighted ones. (Cf. also [**7**, Lemma 3.8].)

LEMMA 7.14. *Let E be any geometric tensor bundle over M, and let $P = \nabla^*\nabla + \mathscr{K}$ be a Laplace operator acting on sections of E. Suppose that P satisfies the asymptotic estimate*

$$(u, Pu) \gtrsim \lambda \|u\|^2 \qquad (7.11)$$

for some $\lambda > 0$. If $|\delta|^2 < \lambda$, there is a compact set $K \subset M$ such that the following weighted estimate holds for all $u \in C_c^\infty(M \smallsetminus K; E)$:

$$\|u\|_{0,2,\delta} \leq C\|Pu\|_{0,2,\delta}. \qquad (7.12)$$

PROOF. If $u \in C_c^\infty(M; E)$, then
$$\begin{aligned}(\rho^{-\delta}u, \nabla^*\nabla(\rho^{-\delta}u)) &= \|\nabla(\rho^{-\delta}u)\|_g^2 \\ &= \int_M \left(\rho^{-2\delta}|\nabla u|_g^2 - 2\delta\rho^{-2\delta}\left\langle \nabla u, u \otimes \frac{d\rho}{\rho}\right\rangle_g \right. \\ &\quad \left. + \delta^2 \rho^{-2\delta}|u|_g^2 \left|\frac{d\rho}{\rho}\right|_g^2\right) dV_g.\end{aligned} \quad (7.13)$$

By the divergence theorem and the fact that $\Delta|u|_g^2 = 2\langle u, \nabla^*\nabla u\rangle_g - 2|\nabla u|_g^2$,
$$\begin{aligned}0 &= \frac{1}{2}\int_M d^*\left(\rho^{-2\delta}d|u|_g^2\right) dV_g \\ &= \int_M \rho^{-2\delta}\langle u, \nabla^*\nabla u\rangle_g - \rho^{-2\delta}|\nabla u|_g^2 + 2\delta\rho^{-2\delta}\left\langle \nabla u, u \otimes \frac{d\rho}{\rho}\right\rangle_g dV_g.\end{aligned}$$

Substituting this into (7.13), we obtain
$$(\rho^{-\delta}u, \nabla^*\nabla(\rho^{-\delta}u)) = (\rho^{-\delta}u, \rho^{-\delta}\nabla^*\nabla u) + \delta^2 \int_M \rho^{-2\delta}|u|_g^2 \left|\frac{d\rho}{\rho}\right|_g^2 dV_g. \quad (7.14)$$

Given $\varepsilon > 0$, choose $r > 0$ small enough that
$$(u, Pu) \geq (\lambda - \varepsilon)\|u\|^2 \quad (7.15)$$
whenever u is smooth and compactly supported in A_r, and such that $|d\rho/\rho|_g^2 \leq 1 + \varepsilon/(\delta^2)$ on A_r. Applying (7.15) to $\rho^{-\delta}u$ and using (7.14), we obtain
$$\begin{aligned}(\lambda - \varepsilon)\|\rho^{-\delta}u\|^2 &\leq (\rho^{-\delta}u, (\nabla^*\nabla + \mathscr{K})(\rho^{-\delta}u)) \\ &\leq (\rho^{-\delta}u, \rho^{-\delta}(\nabla^*\nabla + \mathscr{K})u) + (\delta^2 + \varepsilon)\|u\|_{0,2,\delta}^2 \\ &\leq \|u\|_{0,2,\delta}\|Pu\|_{0,2,\delta} + (\delta^2 + \varepsilon)\|u\|_{0,2,\delta}^2,\end{aligned}$$
which implies (7.12) as long as $2\varepsilon < \lambda - \delta^2$. $\square$

With this estimate in hand, it follows immediately from Lemma 4.10 that P is Fredholm on $H_\delta^{k,2}(M;E)$ provided that $|\delta|^2 < \lambda$. If only L^2 results are needed, this provides an exceedingly elementary approach that can be used in many cases.

CHAPTER 8

Einstein Metrics

In this chapter, we will apply the linear theory we have developed so far to prove the existence of Einstein metrics. To do so, we first need to describe a systematic construction of asymptotic solutions to the Einstein equation, which will then be corrected by appealing to an appropriate linear isomorphism theorem and the inverse function theorem. The construction of asymptotic solutions is very similar to the one we described on hyperbolic space in [29], but more delicate because we start with less regularity at the boundary.

Suppose (M, h) is an asymptotically hyperbolic Einstein $(n + 1)$-manifold of class $C^{l,\beta}$, with $l \geq 2$ and $0 < \beta < 1$. Let ρ be a smooth defining function for h, and put $\overline{h} = \rho^2 h$, $\widehat{h} = \overline{h}|_{\partial M}$.

Recall the spaces $C_{(s)}^{k,\alpha}(\overline{M}; E)$ defined in Chapter 3. By definition of the indicial map, if P is a uniformly degenerate operator of order m and $\overline{u} \in C_{(0)}^{m,\alpha}(\overline{M}; E)$, then the behavior of $P(\rho^s \overline{u})$ at the boundary is dominated by $\rho^s I_s(P)\overline{u}$. This in turn follows from the fact that $\rho^{-s}\partial_{i_1}\cdots\partial_{i_j}(\rho^s\overline{u}) = \rho^{-s}\partial_{i_1}\cdots\partial_{i_j}(\rho^s)\overline{u} + o(\rho^{-j})$ in background coordinates whenever $\overline{u} \in C_{(0)}^{j,\alpha}(\overline{M})$. In the construction of our asymptotic solutions, we will need to make similar estimates when $\overline{u}$ has somewhat less regularity. The key is the following lemma.

PROPOSITION 8.1. *Suppose $0 < \alpha < 1$, $0 < k + \alpha \leq l + \beta$, $\delta \in \mathbb{R}$, and s, j are integers satisfying $1 \leq j \leq s < k + \alpha \leq l + \beta$. If $u \in C_{(s)}^{k,\alpha}(\overline{M})$, then for any indices $1 \leq i_1, \ldots, i_j \leq n+1$,*

$$\rho^{-\delta}\partial_{i_1}\cdots\partial_{i_j}(\rho^\delta u) - \rho^{-\delta-s}\partial_{i_1}\cdots\partial_{i_j}(\rho^{\delta+s})u \in C_{(s-j+\alpha)}^{k-j,\alpha}(\overline{M}).$$

PROOF. The proof is by induction on j. For $j = 1$, consider first the case $\delta = 0$, in which case the claim is

$$\partial_i u = s\rho^{-1}(\partial_i \rho) u \mod C_{(s-1+\alpha)}^{k-1,\alpha}(\overline{M}). \tag{8.1}$$

We will prove this claim by induction on k. Observe that $\partial_i u$ and $s\rho^{-1}(\partial_i \rho)u$ are in $C_{(s-1)}^{k-1,\alpha}(\overline{M})$ by parts (e) and (g) of Lemma 3.1. Thus to prove (8.1), it suffices to show that the difference between the two terms is $O(\rho^{s-1+\alpha})$.

For $k = 1$, the only value of s that satisfies the hypotheses is $s = 1$. Since $\rho = \theta^{n+1}$ is a coordinate function, we consider separately the cases $i < n + 1$ and $i = n + 1$. When $i < n+1$ the right-hand side of (8.1) is zero. On the other hand, the left-hand side is in $C_{(0)}^{0,\alpha}(\overline{M})$ and vanishes on ∂M (since u vanishes on ∂M and ∂_i is tangent to ∂M), so it is $O(\rho^\alpha)$. When $i = n+1$, $\rho^{-1}u \in C_{(0)}^{0,\alpha}(\overline{M})$ by Lemma

3.1(g), and by definition of the derivative,

$$\partial_{n+1}u(\theta,0) = \lim_{\rho\to 0}\frac{u(\theta,\rho)-u(\theta,0)}{\rho}$$
$$= \lim_{\rho\to 0}\rho^{-1}u(\theta,\rho).$$

It follows that $\partial_{n+1}u - \rho^{-1}(\partial_{n+1}\rho)u = \partial_{n+1}u - \rho^{-1}u \in C^{0,\alpha}_{(0)}(\overline{M})$ and vanishes on ∂M, so it too is $O(\rho^\alpha)$. This proves (8.1) in the $k=s=1$ case.

Now let $k > 1$, and suppose (8.1) holds for all smaller values of k. Observe that $v = \rho^{1-s}u \in C^{k-s+1,\alpha}_{(1)}(\overline{M})$ by Lemma 3.1(g). Therefore, by the chain rule and the inductive hypothesis,

$$\begin{aligned}\partial_i u &= \partial_i(\rho^{s-1}v) \\ &= (s-1)\rho^{s-2}(\partial_i\rho)v + \rho^{s-1}\partial_i v \\ &= (s-1)\rho^{-1}(\partial_i\rho)u + \rho^{s-1}(\rho^{-1}(\partial_i\rho)v + O(\rho^\alpha)) \\ &= s\rho^{-1}(\partial_i\rho)u + O(\rho^{s-1+\alpha}).\end{aligned}$$

Finally, for $\delta \neq 0$, the product rule gives

$$\begin{aligned}\rho^{-\delta}\partial_i(\rho^\delta u) &= \rho^{-\delta}(\delta\rho^{\delta-1}(\partial_i\rho)u + \rho^\delta\partial_i u) \\ &= \delta\rho^{-1}(\partial_i\rho)u + s\rho^{-1}(\partial_i\rho)u + O(\rho^{s-1+\alpha}) \\ &= \rho^{-\delta-s}\partial_i(\rho^{\delta+s})u + O(\rho^{s-1+\alpha}).\end{aligned}$$

This completes the proof of the $j=1$ step.

Now suppose the proposition is true for some $j \geq 1$. For any $(j+1)$-tuple $(i_1,\ldots,i_{j+1})$, beginning with the induction hypothesis in the form

$$\partial_{i_2}\cdots\partial_{i_{j+1}}(\rho^\delta u) - \rho^{-s}\partial_{i_2}\cdots\partial_{i_{j+1}}(\rho^{\delta+s})u$$
$$\in \rho^\delta C^{k-j,\alpha}_{(s-j+\alpha)}(\overline{M}) \subset C^{k-j,\alpha}_{(\delta+s-j+\alpha)}(\overline{M}),$$

we apply the chain rule to obtain

$$\begin{aligned}\rho^{-\delta}\partial_{i_1}\cdots\partial_{i_{l+1}}(\rho^\delta u) &= \rho^{-\delta}\big(-s\rho^{-s-1}(\partial_{i_1}\rho)\partial_{i_2}\cdots\partial_{i_{l+1}}(\rho^{\delta+s})u \\ &\quad + \rho^{-s}\partial_{i_1}\cdots\partial_{i_{l+1}}(\rho^{\delta+s})u \\ &\quad + \rho^{-s}\partial_{i_2}\cdots\partial_{i_{l+1}}(\rho^{\delta+s})\partial_{i_1}u\big) + O(\rho^{s-j-1+\alpha}) \\ &= \rho^{-\delta-s}\partial_{i_1}\cdots\partial_{i_{l+1}}(\rho^{\delta+s})u + O(\rho^{s-j-1+\alpha}),\end{aligned}$$

where in the last line we have used the induction hypothesis again to evaluate $\partial_{i_1}u$. □

COROLLARY 8.2. *Let* $P\colon C^\infty(M;E) \to C^\infty(M;E)$ *be a self-adjoint, elliptic, geometric partial differential operator of order* $m \leq l$. *Suppose* $0 < \alpha < 1$, $0 < k+\alpha \leq l+\beta$, $\delta \in \mathbb{R}$, *and* $u \in C^{k,\alpha}_{(s)}(\overline{M};E)$, *where* s *is an integer satisfying* $1 \leq s < k+\alpha \leq l+\beta$. *Then*

$$\rho^{-\delta-s}P(\rho^\delta u)|_{\partial M} = I_{\delta+s}(P)\widehat{u}, \tag{8.2}$$

where $\widehat{u} = \rho^{-s}u|_{\partial M}$.

PROOF. Write Pu in background coordinates as

$$Pu = \sum_{0 \leq j \leq m} \sum_{i_1,\ldots,i_j} \rho^j a^{i_1,\ldots,i_j} \partial_{i_1} \cdots \partial_{i_j} u,$$

where u is vector-valued and the coefficient functions $a^{i_1,\ldots,i_j}$ are matrix-valued. If $\overline{u} \in C^{m,\alpha}_{(0)}(M;E)$, then an easy computation shows that

$$\rho^{-\delta-s} P(\rho^{\delta+s}\overline{u}) = \rho^{-\delta-s} \sum_{0 \leq j \leq m} \sum_{i_1,\ldots,i_j} \rho^j a^{i_1,\ldots,i_j} \partial_{i_1} \cdots \partial_{i_j}(\rho^{\delta+s})\overline{u} + o(1),$$

which implies that

$$I_{\delta+s}(P)\widehat{u} = \sum_{0 \leq j \leq m} \sum_{i_1,\ldots,i_j} (\rho^{-\delta-s} \rho^j a^{i_1,\ldots,i_j} \partial_{i_1} \cdots \partial_{i_j}(\rho^{\delta+s}))|_{\partial M} \widehat{u}.$$

On the other hand, Proposition 8.1 shows that

$$\rho^{-\delta-s} P(\rho^s u) = \sum_{0 \leq j \leq m} \sum_{i_1,\ldots,i_j} \rho^{-\delta-s} \rho^j a^{i_1,\ldots,i_j} \partial_{i_1} \cdots \partial_{i_j}(\rho^{\delta+s})(\rho^{-s} u) + O(\rho^\alpha),$$

which proves the result. □

We will need an extension operator from tensor fields on ∂M to $C^{l,\beta}$ tensor fields on $\overline{M}$. In [29], we did this by parallel translating along $\overline{h}$-geodesics normal to ∂M. However, in our present circumstances, because $\overline{h}$ may not be smooth up to the boundary, this parallel translation would lose regularity. Instead, we define our extension operator as follows.

Let $\pi \colon T\overline{M}|_{\partial M} \to T\partial M$ denote the $\overline{h}$-orthogonal projection, which is clearly a $C^{l,\beta}$ bundle map over ∂M. Given any $C^{l,\beta}$ 2-tensor field v on ∂M, lifting by π yields a $C^{l,\beta}$ section $\pi^* v$ of $\Sigma^2 \overline{M}|_{\partial M}$:

$$\pi^* v(X,Y) = v(\pi X, \pi Y).$$

Define a linear map $E \colon C^{l,\beta}(\partial M; \Sigma^2 \partial M) \to C^{l,\beta}_{(0)}(\overline{M}; \Sigma^2 \overline{M})$ by letting $E(\widehat{u}) = \varphi \Pi(\pi^* u)$, where φ is a fixed cutoff function that is equal to 1 along ∂M and is supported in a small collar neighborhood of the boundary, and Π denotes parallel translation along normal geodesics with respect to some smooth background metric.

For any $C^{l,\beta}$ Riemannian metric $\widehat{g}$ on ∂M, we define a metric $T(\widehat{g})$ on M by

$$T(\widehat{g}) = h + \rho^{-2} E(\widehat{g} - \widehat{h}). \tag{8.3}$$

Then $T \colon C^{l,\beta}(\partial M; \Sigma^2 \partial M) \to C^{l,\beta}(M; \Sigma^2 M)$ defines a smooth (in fact affine) map of Banach spaces. It is easy to see that if $\widehat{g}$ is sufficiently close to $\widehat{h}$ in the $C^{l,\beta}$ norm, then $T(\widehat{g})$ will be an asymptotically hyperbolic metric on M of class $C^{l,\beta}$, whose conformal infinity is $[\widehat{g}]$. Moreover, since $E(0) = 0$, our construction guarantees that $T(\widehat{h}) = h$.

For higher-order asymptotics, we will use an extension lemma due to L. Andersson and P. Chruściel [8, Lemma 3.3.1]. Translated into our notation, it reads as follows.

LEMMA 8.3. *Suppose $0 \leq \alpha < 1$ and $k \geq 0$. Given an integer s such that $0 \leq s \leq k$ and any $\psi \in C^{k-s,\alpha}(\partial M)$, there exists a function $u \in C^{k,\alpha}_{(s)}(\overline{M})$ such that $(\rho^{-s} u)|_{\partial M} = \psi$.*

The formula [**8**, (3.3.6)] for the derivatives of u makes it clear that the mapping $\psi \mapsto u$ is a bounded linear map from $C^{k-s,\alpha}(\partial M)$ to $C^{k,\alpha}_{(s)}(\overline{M})$.

Recall the operator Q defined by (1.2). The next lemma gives a construction of asymptotic solutions to $Q(g, g_0) = 0$ analogous to Theorem 2.11 of [**29**].

LEMMA 8.4. *Suppose $0 < \beta < 1$, $2 \leq l \leq n-1$, and h is an asymptotically hyperbolic metric on M of class $C^{l,\beta}$. Let $\widehat{g}$ be any metric on ∂M of class $C^{l,\beta}$, and set $g_0 = T(\widehat{g})$. There exists an asymptotically hyperbolic metric g of class $C^{l,\beta}$ on M such that $\rho^2 g|_{\partial M} = \widehat{g}$ and*

$$Q(g, g_0) \in C^{l-2,\beta}_{(l-2+\beta)}(\overline{M}; \Sigma^2 \overline{M}) \subset C^{l-2,\beta}_{l+\beta}(M; \Sigma^2 M). \tag{8.4}$$

The mapping $S \colon C^{l,\beta}(\partial M; \Sigma^2 \partial M) \to \rho^{-2} C^{l,\beta}_{(0)}(\overline{M}; \Sigma^2 \overline{M})$ given by $S(\widehat{g}) = g$ is a smooth map of Banach spaces.

PROOF. Define a nonlinear operator

$$\overline{Q} \colon C^{l,\beta}_{(0)}(\overline{M}; \Sigma^2 \overline{M}) \times C^{l,\beta}_{(0)}(\overline{M}; \Sigma^2 \overline{M}) \to C^{l-2,\beta}_{(0)}(\overline{M}; \Sigma^2 \overline{M})$$

by

$$\overline{Q}(\overline{g}, \overline{g}_0) = \rho^2 Q(\rho^{-2} g, \rho^{-2} g_0).$$

It is clear that (8.4) is equivalent to $\overline{Q}(\rho^2 g, \rho^2 g_0) = O(\rho^{l-2+\beta})$. It follows from formula (2.19) of [**29**] that $\overline{Q}$ has the following expression in background coordinates:

$$\overline{Q}(\overline{g}, \overline{g}_0) = \mathscr{E}^0(\overline{g}, \overline{g}_0) + \rho \mathscr{E}^1(\overline{g}, \overline{g}_0) + \rho^2 \mathscr{E}^2(\overline{g}, \overline{g}_0),$$

where $\mathscr{E}^j$ is a universal polynomial in the components of $\overline{g}$, $\overline{g}_0$, their inverses, and their coordinate derivatives of order less than or equal to j. It follows that $\overline{Q}$ takes its values in

$$C^{l,\beta}_{(0)}(\overline{M}; \Sigma^2 \overline{M}) + \rho C^{l-1,\beta}_{(0)}(\overline{M}; \Sigma^2 \overline{M}) + \rho^2 C^{l-2,\beta}_{(0)}(\overline{M}; \Sigma^2 \overline{M}), \tag{8.5}$$

and is a smooth map of Banach spaces.

We will recursively construct a sequence of metrics $\overline{g}_0, \ldots, \overline{g}_l \in C^{l,\beta}_{(0)}(\overline{M}; \Sigma^2 \overline{M})$ satisfying $\overline{Q}(\overline{g}_k, \overline{g}_0) = o(\rho^k)$. In fact, we will prove a bit more, namely that

$$\overline{Q}(\overline{g}_k, \overline{g}_0) \in \begin{cases} \rho C^{l-1,\beta}_{(0)}(\overline{M}; \Sigma^2 \overline{M}) + \rho^2 C^{l-2,\beta}_{(0)}(\overline{M}; \Sigma^2 \overline{M}), & k = 0, \\ \rho^2 C^{l-2,\beta}_{(k-1)}(\overline{M}; \Sigma^2 \overline{M}), & 1 \leq k \leq l-1, \\ \rho^2 C^{l-2,\beta}_{(l-2+\beta)}(\overline{M}; \Sigma^2 \overline{M}), & k = l. \end{cases} \tag{8.6}$$

Begin with $\overline{g}_0 = \rho^2 T(\widehat{g})$. It follows from Corollary 2.6 of [**29**] that

$$\overline{Q}(\overline{g}_0, \overline{g}_0) = O(\rho).$$

Since the intersection of (8.5) with $O(\rho)$ is

$$C^{l,\beta}_{(1)}(\overline{M}; \Sigma^2 \overline{M}) + \rho C^{l-1,\beta}_{(0)}(\overline{M}; \Sigma^2 \overline{M}) + \rho^2 C^{l-2,\beta}_{(0)}(\overline{M}; \Sigma^2 \overline{M})$$
$$\subset \rho C^{l-1,\beta}_{(0)}(\overline{M}; \Sigma^2 \overline{M}) + \rho^2 C^{l-2,\beta}_{(0)}(\overline{M}; \Sigma^2 \overline{M}),$$

we have (8.6) in the $k = 0$ case.

Assume by induction that for some k, $1 \leq k \leq l$, we have constructed $\overline{g}_{k-1} \in C^{l,\beta}_{(0)}(\overline{M}; \Sigma^2 \overline{M})$ satisfying the analogue of (8.6). Then letting $\widehat{v} = \rho^{-k} \overline{Q}(\overline{g}_{k-1}, \overline{g}_0)|_{\partial M}$, we see from Lemma 3.1(g) that $\widehat{v}$ is a $C^{l-k,\beta}$ section of $T\overline{M}|_{\partial M}$.

8. EINSTEIN METRICS

If $\overline{r} \in C^{l,\beta}_{(k)}(\overline{M}; \Sigma^2 \overline{M})$, the same argument as in the proof of [**29**, Theorem 2.11] shows that

$$\overline{Q}(\overline{g}_{k-1} + \overline{r}, \overline{g}_0) = \overline{Q}(\overline{g}_{k-1}, \overline{g}_0) + \rho^2 (\Delta_L + 2n)(\rho^{-2}\overline{r}) + o(\rho^k). \tag{8.7}$$

By Corollary 8.2, we have

$$\rho^{2-k}(\Delta_L + 2n)(\rho^{-2}\overline{r})|_{\partial M} = I_{k-2}(\Delta_L + 2n)(\rho^{-k}\overline{r})|_{\partial M} + o(1).$$

Because the indicial radius of $\Delta_L + 2n$ is $R = n/2$, $I_s(\Delta_L + 2n)$ is invertible provided $-2 < s < n-2$. Thus as long as $1 \le k \le l \le n-1$, there is a unique $C^{l-k,\beta}$ tensor field ψ along ∂M such that $I_{k-2}(\Delta_L + 2n)\psi = -\widehat{v}$. By Lemma 8.3, there is a tensor field $\overline{r} \in C^{l,\beta}_{(k)}(\overline{M}; \Sigma^2 \overline{M})$ such that $(\rho^{-k}\overline{r})|_{\partial M} = \psi$. Inserting this back into (8.7), we conclude that $\overline{Q}(\overline{g}_{k-1} + \overline{r}, \overline{g}_0) = o(\rho^k)$, and thus actually satisfies (8.6). Setting $\overline{g}_k = \overline{g}_{k-1} + \overline{r}$ completes the inductive step.

After the $k = l$ step, we set $g = g_l$, and (8.4) is satisfied. The smoothness of the operator $S: \widehat{g} \mapsto g$ is immediate from the remark following Lemma 8.3 above. □

PROOF OF THEOREM A. The proof follows closely that of Theorem 4.1 in [**29**]. We define an open subset $\mathscr{B} \subset C^{l,\beta}(\partial M; \Sigma^2 \partial M) \times C^{l,\beta}_{l+\beta}(M; \Sigma^2 M)$ by

$$\mathscr{B} = \{(\widehat{g}, r) : \widehat{g}, T(\widehat{g}), \text{ and } S(\widehat{g}) + r \text{ are all positive definite}\}.$$

Define a map $\mathscr{Q} : \mathscr{B} \to C^{l,\beta}(\partial M; \Sigma^2 \partial M) \times C^{l-2,\beta}_{l+\beta}(M; \Sigma^2 M)$ by

$$\mathscr{Q}(\widehat{g}, r) = (\widehat{g}, Q(S(\widehat{g}) + r, T(\widehat{g}))),$$

where T is defined in (8.3) and S in Lemma 8.4. It follows just as in [**29**] that $\mathscr{Q}$ is a smooth map of Banach spaces.

Since $T(\widehat{h}) = S(\widehat{h}) = h$ and h is an Einstein metric, $\mathscr{Q}(\widehat{h}, 0) = (\widehat{h}, 0)$. As is shown in [**29**], the linearization of $\mathscr{Q}$ about $(\widehat{h}, 0)$ is the linear map $D\mathscr{Q}_{(\widehat{h},0)}$ from $C^{l,\beta}(\partial M; \Sigma^2 \partial M) \times C^{l,\beta}_{l+\beta}(M; \Sigma^2 M)$ to $C^{l,\beta}(\partial M; \Sigma^2 \partial M) \times C^{l-2,\beta}_{l+\beta}(M; \Sigma^2 M)$ given by

$$D\mathscr{Q}_{(\widehat{h},0)}(\widehat{q}, r) = (\widehat{q}, D_1 Q_{(h,h)}(DS_{\widehat{h}}\widehat{q} + r) + D_2 Q_{(h,h)}(DT_{\widehat{h}}\widehat{q}))$$
$$= (\widehat{q}, (\Delta_L + 2n)r + K\widehat{q}),$$

where

$$K\widehat{q} = D_1 Q_{(h,h)}(DS_{\widehat{h}}\widehat{q}) + D_2 Q_{(h,h)}(DT_{\widehat{h}}\widehat{q}).$$

Because $\Delta_L + 2n$ preserves the splitting $\Sigma^2 M = \Sigma^2_0 M \oplus \mathbb{R}g$, to show that it satisfies the hypotheses of Theorem C, we need to show that it has trivial L^2 kernel on each of these bundles separately. Since $\overset{\circ}{Rm}(g) = \overset{\circ}{Rc}(g)$, the action of Δ_L on sections of $\mathbb{R}g$ is just given by the scalar Laplacian:

$$\Delta_L(ug) = (\nabla^*\nabla u)g.$$

Since $2n > 0$, it follows easily from integration by parts that $\nabla^*\nabla u + 2n$ acting on scalar functions has trivial L^2 kernel. Thus the assumption that $\Delta_L + 2n$ has trivial L^2 kernel on $\Sigma^2_0 M$ is sufficient to allow us to apply the results of Theorem C. In particular, $\Delta_L + 2n: C^{l,\beta}_\delta(M; \Sigma^2 M) \to C^{l-2,\beta}_\delta(M; \Sigma^2 M)$ is an isomorphism for $|\delta - n/2| < n/2$, which is to say for $0 < \delta < n$. This is true for $\delta = l + \beta$, so the linearization of $\mathscr{Q}$ has a bounded inverse given by

$$(D\mathscr{Q}_{(\widehat{h},0)})^{-1}(\widehat{w}, v) = (\widehat{w}, (\Delta_L + 2n)^{-1}(v - K\widehat{w})).$$

Therefore, by the Banach space inverse function theorem, there is a neighborhood of $(\widehat{h}, 0)$ on which $\mathcal{Q}$ has a smooth inverse. In particular, for $\widehat{g}$ sufficiently $C^{l,\beta}$ close to $\widehat{h}$, there is a solution $r \in C^{l,\beta}_{l+\beta}(M; \Sigma^2 M)$ to $\mathcal{Q}(\widehat{g}, r) = (\widehat{g}, 0)$.

Putting $g = S(\widehat{g}) + r$ and $g_0 = T(\widehat{g})$, we have $Q(g, g_0) = 0$. Moreover, if the $C^{l,\beta}$ neighborhood of $\widehat{h}$ is sufficiently small, then g will be uniformly C^2 close to h, and therefore g will have strictly negative Ricci curvature. By Lemma 2.2 of [**29**], this implies that g is Einstein. Note that $\rho^2 r \in C^{l,\beta}_{l+2+\beta}(M; \Sigma^2 M)$, which is contained in $C^{l,\beta}_{(0)}(\overline{M}; \Sigma^2 \overline{M})$ by Lemma 3.7. Since $\rho^2 S(\widehat{g}) \in C^{l,\beta}_{(0)}(\overline{M}; \Sigma^2 \overline{M})$ by construction, it follows that g is asymptotically hyperbolic of class $C^{l,\beta}$ as claimed.

It remains only to prove that $\Delta_L + 2n$ has trivial L^2 kernel on $\Sigma^2_0 M$ under the assumptions stated in Theorem A. First suppose that h has nonpositive sectional curvature. A simple algebraic argument (see [**13**, Lemma 12.71]) shows that if h is an Einstein metric on an $(n+1)$-manifold with scalar curvature $-n(n+1)$, its Riemann curvature operator $\mathring{R}m$ acting on trace-free symmetric 2-tensors satisfies the following estimate at each point $p \in M$:

$$\langle \mathring{R}m(u), u \rangle_h \leq \left(n + (n-1) K_{\max}(p)\right) |u|^2_h, \tag{8.8}$$

where $K_{\max}(p)$ is the maximum of the sectional curvatures of h at p. (In [**13**], this is attributed to a hard-to-find 1979 paper of T. Fujitani; however, the argument was already given in 1978 by N. Koiso [**35**, Prop. 3.4].) Since the Einstein assumption implies that $\mathring{R}c(u) = -nu$, we have

$$\Delta_L + 2n = \nabla^* \nabla - 2\mathring{R}m = DD^* + D^*D + n - \mathring{R}m$$

(see (7.10)). Therefore, if h has sectional curvatures everywhere bounded above by $-\kappa \leq 0$, for any $u \in C^\infty_c(M; \Sigma^2_0 M)$ we have

$$\begin{aligned}(u, (\Delta_L + 2n)u) &= \|D^* u\|^2 + \|Du\|^2 + n\|u\|^2 - (u, \mathring{R}m(u)) \\ &\geq (u, (n - \mathring{R}m)u) \\ &\geq (n-1)\kappa \|u\|^2.\end{aligned}$$

The same is true for $u \in H^{2,2}(M; \Sigma^2_0 M)$ because $C^\infty_c(M; \Sigma^2_0 M)$ is dense in that space. If $\kappa > 0$, it follows immediately that $\Delta_L + 2n$ has trivial L^2 kernel. On the other hand, if $\kappa = 0$, the sequence of inequalities above implies that if $u \in H^{2,2}(M; \Sigma^2_0 M)$ is a solution to $(\Delta_L + 2n)u = 0$, the nonnegative quantity $\langle u, (n - \mathring{R}m)u \rangle_h$ must vanish identically on M. Since the sectional curvatures of h approach -1 at infinity, there is some compact set $K \subset M$ such that $K_{\max}(p) \leq -1/2$ for $p \in M \smallsetminus K$, and then (8.8) implies that $u \equiv 0$ on $M \smallsetminus K$. Since Δ_L is a Laplace operator, it satisfies the weak unique continuation property [**9**, **47**], and therefore u is identically zero.

Finally, suppose that the conformal infinity $[\widehat{h}]$ has nonnegative Yamabe invariant. Then the result of [**38**] shows that the Laplacian satisfies the following L^2 estimate for smooth, compactly supported scalar functions u:

$$(u, \nabla^* \nabla u) \geq \frac{n^2}{4} \|u\|^2.$$

(The proof in [**38**] required h to have a $C^{3,\alpha}$ conformal compactification. However, using Lemma 3.3.1 of [**8**], it is easy to reduce that to $C^{2,\alpha}$. See also [**57**] for a

8. EINSTEIN METRICS

different proof.) By Lemma 7.7, therefore, the same is true when u is a smooth, compactly supported tensor field, and by continuity for all $u \in H^{2,2}(M; \Sigma_0^2 M)$.

Suppose h has sectional curvatures bounded above by $(n^2 - 8n)/(8n - 8)$. Then (8.8) gives

$$2\langle u, \mathring{Rm}(u)\rangle_h \leq 2\left(n + \frac{(n-1)(n^2 - 8n)}{8n - 8}\right) |u|_h^2 = \frac{n^2}{4}|u|_h^2.$$

If $u \in L^2(M; \Sigma_0^2 M)$ is a solution to $(\Delta_L + 2n)u = 0$, therefore,

$$\begin{aligned}
0 &= (u, (\Delta_L + 2n)u) \\
&= (u, (\nabla^*\nabla - 2\mathring{Rm})u) \\
&= \|\nabla u\|^2 - \frac{n^2}{4}\|u\|^2 + \left(u, (n^2/4 - 2\mathring{Rm})u\right) \\
&\geq \left(u, (n^2/4 - 2\mathring{Rm})u\right) \\
&\geq 0.
\end{aligned}$$

It follows as before that the nonnegative function $\langle u, (n^2/4 - 2\mathring{Rm})u\rangle_h$ must be identically zero, and since the operator $(n^2/4 - 2\mathring{Rm})$ is positive definite outside a compact set, u must be identically zero by analytic continuation. $\square$

Bibliography

1. S. Agmon, A. Douglis, and L. Nirenberg, *Estimates near the boundary for solutions of elliptic partial differential equations satisfying general boundary conditions. II*, Comm. Pure Appl. Math. **17** (1964), 35–92.
2. Michael T. Anderson, L^2 *curvature and volume renormalization of AHE metrics on 4-manifolds*, Math. Res. Lett. **8** (2001), 171–188.
3. _____, *Einstein metrics with prescribed conformal infinity on 4-manifolds*, arXiv:math.DG/0105243, 2001/2004.
4. _____, *Boundary regularity, uniqueness and non-uniqueness for AH Einstein metrics on 4-manifolds*, Adv. Math. **179** (2003), 205–249.
5. _____, *Some results on the structure of conformally compact Einstein metrics*, arXiv:math.DG/0402198, 2004/2005.
6. _____, *Geometric aspects of the AdS/CFT correspondence*, AdS/CFT correspondence: Einstein metrics and their conformal boundaries, IRMA Lect. Math. Theor. Phys., vol. 8, Eur. Math. Soc., Zürich, 2005, pp. 1–31.
7. Lars Andersson, *Elliptic systems on manifolds with asymptotically negative curvature*, Indiana Univ. Math. J. **42** (1993), 1359–1388.
8. Lars Andersson and Piotr T. Chruściel, *Solutions of the constraint equations in general relativity satisfying "hyperboloidal boundary conditions"*, Dissertationes Math. (Rozprawy Mat.) **355** (1996), 1–100.
9. N. Aronszajn, *A unique continuation theorem for solutions of elliptic partial differential equations or inequalities of second order*, J. Math. Pures Appl. (9) **36** (1957), 235–249.
10. Thierry Aubin, *Some nonlinear problems in Riemannian geometry*, Springer-Verlag, Berlin, 1998.
11. Toby N. Bailey, Michael G. Eastwood, and C. Robin Graham, *Invariant theory for conformal and CR geometry*, Ann. of Math. (2) **139** (1994), 491–552.
12. Robert Bartnik, *The mass of an asymptotically flat manifold*, Comm. Pure Appl. Math. **39** (1986), 661–693.
13. Arthur L. Besse, *Einstein manifolds*, Springer-Verlag, Berlin, 1987.
14. Olivier Biquard, *Einstein deformations of hyperbolic metrics*, Surveys in differential geometry: essays on Einstein manifolds, Int. Press, Boston, MA, 1999, pp. 235–246.
15. _____, *Métriques d'Einstein asymptotiquement symétriques*, Astérisque **265** (2000), vi+109.
16. _____, *Métriques autoduales sur la boule*, Invent. Math. **148** (2002), 545–607.
17. Garrett Birkhoff and Gian-Carlo Rota, *Ordinary differential equations*, fourth ed., John Wiley & Sons Inc., New York, 1989.
18. Mingliang Cai and Gregory J. Galloway, *Boundaries of zero scalar curvature in the AdS/CFT correspondence*, Adv. Theor. Math. Phys. **3** (1999), 1769–1783 (2000), arXiv:hep-th/0003046.
19. Isaac Chavel, *Riemannian geometry—a modern introduction*, Cambridge University Press, Cambridge, 1993.
20. S. Y. Cheng and S. T. Yau, *Differential equations on Riemannian manifolds and their geometric applications*, Comm. Pure Appl. Math. **28** (1975), 333–354.
21. Piotr T. Chruściel, Erwann Delay, John M. Lee, and Dale N. Skinner, *Boundary regularity of conformally compact Einstein metrics*, arXiv:math.DG/0401386, to appear in J. Differential Geom., 2005.
22. Erwann Delay, *Étude locale d'opérateurs de courbure sur l'espace hyperbolique*, J. Math. Pures Appl. (9) **78** (1999), 389–430.
23. _____, *Essential spectrum of the Lichnerowicz Laplacian on two tensors on asymptotically hyperbolic manifolds*, J. Geom. Phys. **43** (2002), 33–44.

24. Erwann Delay and Marc Herzlich, *Ricci curvature in the neighborhood of rank-one symmetric spaces*, J. Geom. Anal. **11** (2001), 573–588.
25. Harold Donnelly and Frederico Xavier, *On the differential form spectrum of negatively curved Riemannian manifolds*, Amer. J. Math. **106** (1984), 169–185.
26. Avron Douglis and Louis Nirenberg, *Interior estimates for elliptic systems of partial differential equations*, Comm. Pure Appl. Math. **8** (1955), 503–538.
27. Arthur Erdélyi, Wilhelm Magnus, Fritz Oberhettinger, and Francesco G. Tricomi, *Higher transcendental functions, Vol. I*, McGraw-Hill Book Company, Inc., New York-Toronto-London, 1953, Based, in part, on notes left by Harry Bateman.
28. Charles Fefferman and C. Robin Graham, *Conformal invariants*, Astérisque (1985), 95–116, The mathematical heritage of Élie Cartan (Lyon, 1984).
29. C. Robin Graham and John M. Lee, *Einstein metrics with prescribed conformal infinity on the ball*, Adv. Math. **87** (1991), 186–225.
30. C. Robin Graham and Edward Witten, *Conformal anomaly of submanifold observables in AdS/CFT correspondence*, Nuclear Phys. B **546** (1999), 52–64.
31. N. J. Hitchin, *Twistor spaces, Einstein metrics and isomonodromic deformations*, J. Differential Geom. **42** (1995), 30–112.
32. _____, *Einstein metrics and the eta-invariant*, Boll. Un. Mat. Ital. B (7) **11** (1997), 95–105.
33. James Isenberg and Jiseong Park, *Asymptotically hyperbolic non-constant mean curvature solutions of the Einstein constraint equations*, Classical Quantum Gravity **14** (1997), A189–A201, Geometry and physics.
34. Tosio Kato, *Perturbation theory for linear operators*, Classics in Mathematics, Springer-Verlag, Berlin, 1995, Reprint of the 1980 edition.
35. Norihito Koiso, *Nondeformability of Einstein metrics*, Osaka J. Math. **15** (1978), 419–433.
36. Robert Lauter, Bertrand Monthubert, and Victor Nistor, *Pseudodifferential analysis on continuous family groupoids*, Doc. Math. **5** (2000), 625–655 (electronic).
37. C. R. LeBrun, *$\mathcal{H}$-space with a cosmological constant*, Proc. Roy. Soc. London Ser. A **380** (1982), 171–185.
38. John M. Lee, *The spectrum of an asymptotically hyperbolic Einstein manifold*, Comm. Anal. Geom. **3** (1995), 253–271.
39. Rafe Mazzeo, *Hodge cohomology of negatively curved manifolds*, Ph.D. thesis, MIT, 1986.
40. _____, *The Hodge cohomology of a conformally compact metric*, J. Differential Geom. **28** (1988), 309–339.
41. _____, *Elliptic theory of differential edge operators. I*, Comm. Partial Differential Equations **16** (1991), 1615–1664.
42. Rafe Mazzeo and Richard B. Melrose, *Meromorphic extension of the resolvent on complete spaces with asymptotically constant negative curvature*, J. Funct. Anal. **75** (1987), 260–310.
43. H. P. McKean, *An upper bound to the spectrum of Δ on a manifold of negative curvature*, J. Differential Geometry **4** (1970), 359–366.
44. Richard B. Melrose, *Transformation of boundary problems*, Acta Math. **147** (1981), 149–236.
45. Richard B. Melrose and Gerardo Mendoza, *Elliptic pseudodifferential operators of totally characteristic type*, MSRI preprint, 1983.
46. Maung Min-Oo, *Scalar curvature rigidity of asymptotically hyperbolic spin manifolds*, Math. Ann. **285** (1989), 527–539.
47. Gen Nakamura, Gunther Uhlmann, and Jenn-Nan Wang, *Unique continuation property for elliptic systems and crack determination in anisotropic elasticity*, Partial differential equations and inverse problems, Contemp. Math., vol. 362, Amer. Math. Soc., Providence, RI, 2004, pp. 321–338.
48. Louis Nirenberg, *Estimates and existence of solutions of elliptic equations*, Comm. Pure Appl. Math. **9** (1956), 509–529.
49. Jiseong Park, *Non-constant mean curvature hyperboloidal solutions of the Einstein constraint equations*, Ph.D. thesis, University of Oregon, 1996.
50. H. Pedersen, *Einstein metrics, spinning top motions and monopoles*, Math. Ann. **274** (1986), 35–59.
51. R. Penrose, *Structure of space-time*, Battelle Rencontres: 1967 Lectures in Mathematics and Physics (Cecile M. DeWitt and John A. Wheeler, eds.), W. A. Benjamin Inc., New York, 1998, pp. 121–235.

52. Jens Lyng Petersen, *Introduction to the Maldacena conjecture on AdS/CFT*, Int. J. Modern Phys. A **14** (1999), 3597–3672.
53. Johan Råde, *Elliptic uniformly degenerate operators*, Chalmers University of Technology (Göteburg University) preprint #1998-19, 1998.
54. John Charles Roth, *Perturbations of Kähler-Einstein metrics*, Ph.D. thesis, University of Washington, 1999.
55. Peter Stredder, *Natural differential operators on Riemannian manifolds and representations of the orthogonal and special orthogonal groups*, J. Differential Geometry **10** (1975), 647–660.
56. Dennis Sullivan, *Related aspects of positivity in Riemannian geometry*, J. Differential Geom. **25** (1987), 327–351.
57. Xiaodong Wang, *A new proof of Lee's theorem on the spectrum of conformally compact Einstein manifolds*, Comm. Anal. Geom. **10** (2002), 647–651.
58. Edward Witten, *Supersymmetry and Morse theory*, J. Differential Geom. **17** (1982), 661–692 (1983).
59. ———, *Anti de Sitter space and holography*, Adv. Theor. Math. Phys. **2** (1998), 253–291, arXiv:hep-th/9802150.
60. Hung Hsi Wu, *The Bochner technique in differential geometry*, Math. Rep. **3** (1988), i–xii and 289–538.

Editorial Information

To be published in the *Memoirs*, a paper must be correct, new, nontrivial, and significant. Further, it must be well written and of interest to a substantial number of mathematicians. Piecemeal results, such as an inconclusive step toward an unproved major theorem or a minor variation on a known result, are in general not acceptable for publication. Papers appearing in *Memoirs* are generally at least 80 and not more than 200 published pages in length. Papers less than 80 or more than 200 published pages require the approval of the Managing Editor of the Transactions/Memoirs Editorial Board.

As of May 31, 2006, the backlog for this journal was approximately 11 volumes. This estimate is the result of dividing the number of manuscripts for this journal in the Providence office that have not yet gone to the printer on the above date by the average number of monographs per volume over the previous twelve months, reduced by the number of volumes published in four months (the time necessary for preparing a volume for the printer). (There are 6 volumes per year, each containing at least 4 numbers.)

A Consent to Publish and Copyright Agreement is required before a paper will be published in the *Memoirs*. After a paper is accepted for publication, the Providence office will send a Consent to Publish and Copyright Agreement to all authors of the paper. By submitting a paper to the *Memoirs*, authors certify that the results have not been submitted to nor are they under consideration for publication by another journal, conference proceedings, or similar publication.

Information for Authors

Memoirs are printed from camera copy fully prepared by the author. This means that the finished book will look exactly like the copy submitted.

The paper must contain a *descriptive title* and an *abstract* that summarizes the article in language suitable for workers in the general field (algebra, analysis, etc.). The *descriptive title* should be short, but informative; useless or vague phrases such as "some remarks about" or "concerning" should be avoided. The *abstract* should be at least one complete sentence, and at most 300 words. Included with the footnotes to the paper should be the 2000 *Mathematics Subject Classification* representing the primary and secondary subjects of the article. The classifications are accessible from www.ams.org/msc/. The list of classifications is also available in print starting with the 1999 annual index of *Mathematical Reviews*. The Mathematics Subject Classification footnote may be followed by a list of *key words and phrases* describing the subject matter of the article and taken from it. Journal abbreviations used in bibliographies are listed in the latest *Mathematical Reviews* annual index. The series abbreviations are also accessible from www.ams.org/publications/. To help in preparing and verifying references, the AMS offers MR Lookup, a Reference Tool for Linking, at www.ams.org/mrlookup/. When the manuscript is submitted, authors should supply the editor with electronic addresses if available. These will be printed after the postal address at the end of the article.

Electronically prepared manuscripts. The AMS encourages electronically prepared manuscripts, with a strong preference for $\mathcal{AMS}$-LaTeX. To this end, the Society has prepared $\mathcal{AMS}$-LaTeX author packages for each AMS publication. Author packages include instructions for preparing electronic manuscripts, the *AMS Author Handbook*, samples, and a style file that generates the particular design specifications of that publication series. Though $\mathcal{AMS}$-LaTeX is the highly preferred format of TeX, author packages are also available in $\mathcal{AMS}$-TeX.

Authors may retrieve an author package from e-MATH starting from www.ams.org/tex/ or via FTP to ftp.ams.org (login as anonymous, enter username as password, and type cd pub/author-info). The *AMS Author Handbook* and the *Instruction Manual* are available in PDF format following the author packages link from www.ams.org/tex/. The author package can also be obtained free of charge by sending email to `tech-support@ams.org` (Internet) or from the Publication Division, American Mathematical Society, 201 Charles St., Providence, RI 02904-2294, USA. When requesting an author package, please specify $\mathcal{AMS}$-LaTeX or $\mathcal{AMS}$-TeX and the publication in which your paper will appear. Please be sure to include your complete mailing address.

Sending electronic files. After acceptance, the source file(s) should be sent to the Providence office (this includes any TeX source file, any graphics files, and the DVI or PostScript file).

Before sending the source file, be sure you have proofread your paper carefully. The files you send must be the EXACT files used to generate the proof copy that was accepted for publication. For all publications, authors are required to send a printed copy of their paper, which exactly matches the copy approved for publication, along with any graphics that will appear in the paper.

TeX files may be submitted by email, FTP, or on diskette. The DVI file(s) and PostScript files should be submitted only by FTP or on diskette unless they are encoded properly to submit through email. (DVI files are binary and PostScript files tend to be very large.)

Electronically prepared manuscripts can be sent via email to `pub-submit@ams.org` (Internet). The subject line of the message should include the publication code to identify it as a Memoir. TeX source files, DVI files, and PostScript files can be transferred over the Internet by FTP to the Internet node `e-math.ams.org` (130.44.1.100).

Electronic graphics. Comprehensive instructions on preparing graphics are available at `www.ams.org/jourhtml/graphics.html`. A few of the major requirements are given here.

Submit files for graphics as EPS (Encapsulated PostScript) files. This includes graphics originated via a graphics application as well as scanned photographs or other computer-generated images. If this is not possible, TIFF files are acceptable as long as they can be opened in Adobe Photoshop or Illustrator. No matter what method was used to produce the graphic, it is necessary to provide a paper copy to the AMS.

Authors using graphics packages for the creation of electronic art should also avoid the use of any lines thinner than 0.5 points in width. Many graphics packages allow the user to specify a "hairline" for a very thin line. Hairlines often look acceptable when proofed on a typical laser printer. However, when produced on a high-resolution laser imagesetter, hairlines become nearly invisible and will be lost entirely in the final printing process.

Screens should be set to values between 15% and 85%. Screens which fall outside of this range are too light or too dark to print correctly. Variations of screens within a graphic should be no less than 10%.

Inquiries. Any inquiries concerning a paper that has been accepted for publication should be sent directly to the Electronic Prepress Department, American Mathematical Society, 201 Charles St., Providence, RI 02904, USA.

Editors

This journal is designed particularly for long research papers, normally at least 80 pages in length, and groups of cognate papers in pure and applied mathematics. Papers intended for publication in the *Memoirs* should be addressed to one of the following editors. In principle the Memoirs welcomes electronic submissions, and some of the editors, those whose names appear below with an asterisk (*), have indicated that they prefer them. However, editors reserve the right to request hard copies after papers have been submitted electronically. Authors are advised to make preliminary email inquiries to editors about whether they are likely to be able to handle submissions in a particular electronic form.

***Algebra** to ALEXANDER KLESHCHEV, Department of Mathematics, University of Oregon, Eugene, OR 97403-1222; email: ams@noether.uoregon.edu

Algebra and its application to MINA TEICHER, Emmy Noether Research Institute for Mathematics, Bar-Ilan University, Ramat-Gan 52900, Israel; email: teicher@macs.biu.ac.il

Algebraic geometry to DAN ABRAMOVICH, Department of Mathematics, Brown University, Box 1917, Providence, RI 02912; email: amsedit@math.brown.edu

***Algebraic number theory** to V. KUMAR MURTY, Department of Mathematics, University of Toronto, 100 St. George Street, Toronto, ON M5S 1A1, Canada; email: murty@math.toronto.edu

***Algebraic topology** to ALEJANDRO ADEM, Department of Mathematics, University of British Columbia, Room 121, 1984 Mathematics Road, Vancouver, British Columbia, Canada V6T 1Z2; email: adem@math.ubc.ca

***Combinatorics** to JOHN R. STEMBRIDGE, Department of Mathematics, University of Michigan, Ann Arbor, Michigan 48109-1109; email: FRS@umich.edu

Complex analysis and harmonic analysis to ALEXANDER NAGEL, Department of Mathematics, University of Wisconsin, 480 Lincoln Drive, Madison, WI 53706-1313; email: nagel@math.wisc.edu

***Differential geometry and global analysis** to LISA C. JEFFREY, Department of Mathematics, University of Toronto, 100 St. George St., Toronto, ON Canada M5S 3G3; email: jeffrey@math.toronto.edu

Dynamical systems and ergodic theory to AMIE WILKINSON, Department of Mathematics, Northwestern University, 2033 Sheridan Road, Evanston, IL 60208-2730; email: transactions@math.northwestern.edu

***Functional analysis and operator algebras** to MARIUS DADARLAT, Department of Mathematics, Purdue University, 150 N. University St., West Lafayette, IN 47907-2067; email: mdd@math.purdue.edu

***Geometric analysis** to TOBIAS COLDING, Courant Institute, New York University, 251 Mercer St., New York, NY 10012; email: traneditor@cims.nyu.edu

***Geometric analysis** to MLADEN BESTVINA, Department of Mathematics, University of Utah, 155 South 1400 East, JWB 233, Salt Lake City, Utah 84112-0090; email: bestvina@math.utah.edu

Harmonic analysis, representation theory, and Lie theory to ROBERT J. STANTON, Department of Mathematics, The Ohio State University, 231 West 18th Avenue, Columbus, OH 43210-1174; email: stanton@math.ohio-state.edu

***Logic** to STEFFEN LEMPP, Department of Mathematics, University of Wisconsin, 480 Lincoln Drive, Madison, Wisconsin 53706-1388; email: lempp@math.wisc.edu

***Ordinary differential equations, and applied mathematics** to PETER W. BATES, Department of Mathematics, Michigan State University, East Lansing, MI 48824-1027; email: bates@math.msu.edu

***Partial differential equations** to GUSTAVO PONCE, Department of Mathematics, South Hall, Room 6607, University of California, Santa Barbara, CA 93106; email: ponce@math.ucsb.edu

***Probability and statistics** to KRZYSZTOF BURDZY, Department of Mathematics, University of Washington, Box 354350, Seattle, Washington 98195-4350; email: burdzy@math.washington.edu

***Real analysis and partial differential equations** to DANIEL TATARU, Department of Mathematics, University of California, Berkeley, Berkeley, CA 94720; email: tataru@math.berkeley.edu

All other communications to the editors should be addressed to the Managing Editor, ROBERT GURALNICK, Department of Mathematics, University of Southern California, Los Angeles, CA 90089-1113; email: guralnic@math.usc.edu.

Titles in This Series

864 **John M. Lee,** Fredholm operators and Einstein metrics on conformally compact manifolds, 2006

863 **M. Lübke and A. Teleman,** The universal Kobayashi-Hitchin correspondence on Hermitian manifolds, 2006

862 **Alberto Canonaco,** The Beilinson complex and canonical rings of irregular surfaces, 2006

861 **Leon A. Takhtajan and Lee-Peng Teo,** Weil-Petersson metric on the universal Teichmüller space, 2006

860 **Thomas M. Fiore,** Pseudo limits, biadjoints and pseudo algebras: Categorical foundations of conformal field theory, 2006

859 **N. Arcozzi, R. Rochberg, and E. Sawyer,** Carleson measures and interpolating sequences for Besov spaces on complex balls, 2006

858 **Enrico Valdinoci, Berardino Sciunzi, and Vasile Ovidiu Savin,** Flat level set regularity of p-Laplace phase transitions, 2006

857 **Donatella Danielli, Nocola Garofalo, and Duy-Minh Nhieu,** Non-doubling Ahlfors measures, perimeter measures, and the characterization of the trace spaces of Sobolev functions in Carnot-Carathéodory spaces, 2006

856 **Vladimir Bolotnikov and Harry Dym,** On boundary interpolation for matrix valued Schur functions, 2006

855 **Yevgenia Kashina, Yorck Sommerhäuser, and Yongchang Zhu,** On higher Frobenius-Schur indicators, 2006

854 **Noam Greenberg,** The role of true finiteness in the admissible recursively enumerable degrees, 2006

853 **Joachim Krieger,** Stability of spherically symmetric wave maps, 2006

852 **Viorel Barbu, Irena Lasiecka, and Roberto Triggiani,** Tangential boundary stabilization of Navier-Stokes equations, 2006

851 **Jie Wu,** On maps from loop suspensions to loop spaces and the shuffle relations on the Cohen groups, 2006

850 **Siegfried Echterhoff, S. Kaliszewski, John Quigg, and Iain Raeburn,** A categorical approach to imprimitivity theorems for C^*-dynamical systems, 2006

849 **Katsuhiko Kuribayashi, Mamoru Mimura, and Tetsu Nishimoto,** Twisted tensor products related to the cohomology of the classifying spaces of loop groups, 2006

848 **Bob Oliver,** Equivalences of classifying spaces completed at the prime two, 2006

847 **Eric T. Sawyer and Richard L. Wheeden,** Hölder continuity of weak solutions to subelliptic equations with rough coefficients, 2006

846 **Victor Beresnevich, Detta Dickinson, and Sanju Velani,** Measure theoretic laws for lim–sup sets, 2006

845 **Ehud Friedgut, Vojtech Rödl, Andrzej Ruciński, and Prasad V. Tetali,** A Sharp threshold for random graphs with a monochromatic triangle in every edge coloring, 2006

844 **Amadeu Delshams, Rafael de la Llave, and Tere M. Seara,** A geometric mechanism for diffusion in Hamiltonian systems overcoming the large gap problem: Heuristics and rigorous verification on a model, 2006

843 **Denis V. Osin,** Relatively hyperbolic groups: Intrinsic geometry, algebraic properties, and algorithmic problems, 2006

842 **David P. Blecher and Vrej Zarikian,** The calculus of one-sided M-ideals and multipliers in operator spaces, 2006

841 **Enrique Artal Bartolo, Pierrette Cassou-Noguès, Ignacio Luengo, and Alejandro Melle Hernández,** Quasi-ordinary power series and their zeta functions, 2005

840 **Sławomir Kołodziej,** The complex Monge-Ampère equation and pluripotential theory, 2005

TITLES IN THIS SERIES

- 839 Mihai Ciucu, A random tiling model for two dimensional electrostatics, 2005
- 838 V. Jurdjevic, Integrable Hamiltonian systems on complex Lie groups, 2005
- 837 Joseph A. Ball and Victor Vinnikov, Lax-Phillips scattering and conservative linear systems: A Cuntz-algebra multidimensional setting, 2005
- 836 H. G. Dales and A. T.-M. Lau, The second duals of Beurling algbras, 2005
- 835 Kiyoshi Igusa, Higher complex torsion and the framing principle, 2005
- 834 Kenichi Ohshika, Kleinian groups which are limits of geometrically finite groups, 2005
- 833 Greg Hjorth and Alexander S. Kechris, Rigidity theorems for actions of product groups and countable Borel equivalence relations, 2005
- 832 Lee Klingler and Lawrence S. Levy, Representation type of commutative Noetherian rings III: Global wildness and tameness, 2005
- 831 K. R. Goodearl and F. Wehrung, The complete dimension theory of partially ordered systems with equivalence and orthogonality, 2005
- 830 Jason Fulman, Peter M. Neumann, and Cheryl E. Praeger, A generating function approach to the enumeration of matrices in classical groups over finite fields, 2005
- 829 S. G. Bobkov and B. Zegarlinski, Entropy bounds and isoperimetry, 2005
- 828 Joel Berman and Paweł M. Idziak, Generative complexity in algebra, 2005
- 827 Trevor A. Welsh, Fermionic expressions for minimal model Virasoro characters, 2005
- 826 Guy Métivier and Kevin Zumbrun, Large viscous boundary layers for noncharacteristic nonlinear hyperbolic problems, 2005
- 825 Yaozhong Hu, Integral transformations and anticipative calculus for fractional Brownian motions, 2005
- 824 Luen-Chau Li and Serge Parmentier, On dynamical Poisson groupoids I, 2005
- 823 Claus Mokler, An analogue of a reductive algebraic monoid whose unit group is a Kac-Moody group, 2005
- 822 Stefano Pigola, Marco Rigoli, and Alberto G. Setti, Maximum principles on Riemannian manifolds and applications, 2005
- 821 Nicole Bopp and Hubert Rubenthaler, Local zeta functions attached to the minimal spherical series for a class of symmetric spaces, 2005
- 820 Vadim A. Kaimanovich and Mikhail Lyubich, Conformal and harmonic measures on laminations associated with rational maps, 2005
- 819 F. Andreatta and E. Z. Goren, Hilbert modular forms: Mod p and p-adic aspects, 2005
- 818 Tom De Medts, An algebraic structure for Moufang quadrangles, 2005
- 817 Javier Fernández de Bobadilla, Moduli spaces of polynomials in two variables, 2005
- 816 Francis Clarke, Necessary conditions in dynamic optimization, 2005
- 815 Martin Bendersky and Donald M. Davis, V_1-periodic homotopy groups of $SO(n)$, 2004
- 814 Johannes Huebschmann, Kähler spaces, nilpotent orbits, and singular reduction, 2004
- 813 Jeff Groah and Blake Temple, Shock-wave solutions of the Einstein equations with perfect fluid sources: Existence and consistency by a locally inertial Glimm scheme, 2004
- 812 Richard D. Canary and Darryl McCullough, Homotopy equivalences of 3-manifolds and deformation theory of Kleinian groups, 2004
- 811 Ottmar Loos and Erhard Neher, Locally finite root systems, 2004
- 810 W. N. Everitt and L. Markus, Infinite dimensional complex symplectic spaces, 2004

For a complete list of titles in this series, visit the
AMS Bookstore at **www.ams.org/bookstore/**.